T0348566

SEISMIC DATA ANALYSIS TECHNIQUES IN HYDROCARBON EXPLORATION

SEISMIC DATA ANALYSIS TECHNIQUES IN HYDROCARBON EXPLORATION

ENWENODE ONAJITE

Research Geophysicist,
Graduate Geoscientists Training Consultant,
Email: jenwenode@yahoo.com

ELSEVIER

AMSTERDAM • BOSTON • HEIDELBERG • LONDON • NEW YORK • OXFORD
PARIS • SAN DIEGO • SAN FRANCISCO • SYDNEY • TOKYO

Elsevier
225 Wyman Street, Waltham, MA 02451, USA
The Boulevard, Langford Lane, Kidlington, Oxford, OX5 1GB, UK
Radarweg 29, PO Box 211, 1000 AE Amsterdam, The Netherlands

Library of Congress Cataloging-in-Publication Data
Onajite, Enwenode.
  Seismic data analysis techniques in hydrocarbon exploration / Enwenode Onajite.
    pages cm
  Includes index.
  ISBN 978-0-12-420023-4
1. Petroleum–Prospecting. 2. Seismic prospecting. I. Title.
  TN271.P4O55 2013
  622'.1828–dc23

                                                                              2013028150

British Library Cataloguing in Publication Data
A catalogue record for this book is available from the British Library

ISBN: 978-0-12-420023-4

For information on all Elsevier publications
visit our website at store.elsevier.com

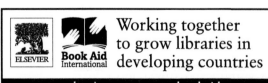

Working together
to grow libraries in
developing countries

www.elsevier.com • www.bookaid.org

Contents

III

SEISMIC DATA INTERPRETATION METHODOLOGY

13. Understanding Seismic Interpretation Methodology

14. Understanding Reflection Coefficient

Acknowledgements

If you want to write a practical handbook on seismic data that would be read by tens of thousands or hopefully millions of geoscientists worldwide, it will definitely take an entire team to achieve such a fit.

The insight and clarity of thought gained, which enabled me to write and build the contents of this industry-oriented book, are not just through my effort and knowledge but also through the wealth of experience and talents of highly skilled industry professionals and resources of many organizations and institutions.

My heartfelt gratitude goes to the following industry professionals who personally guide me with their experiences and knowledge: Dr. Rocco Detomo (Shell E&P, Texas, USA), Dr. James Edet (Total E&P, Nigeria) and Dr. Omu Ugborugbo (Shell E&P, Nigeria). Also, my sincere appreciation is extending to the following individuals for their support and generous contributions:

Alistar R. Brown, Bernard Eromosele, Schlumberger Nigeria

Dr. Adabanija Tunde, a Geologist at Olabisi Olabajo University

Dr. Dorothy Satterfield, Geography, Earth and Environmental Sciences, University of Derby (FEHS)

Dr. Etim D. Uko, a Geophysicist at Rivers State University of Science and Technology, Port Harcourt, Nigeria

Dr. Martin J. Whiteley, Barrisdale Lmt.

Dr. Obiadi Izuchukwu Ignatius, a Geologist at Nnamdi Azikiwe University, Awka, Nigeria

Emo Dayetuawei, Nigeria Association of Petroleum Explorationists, Port Harcourt Branch

Emudianughe Juliet, a Geophysicist at Federal University of Petroleum Resources Effurun (FUPRE)

Ikechukwu Njoku, Geologists

Prahlad Basak, Seismic Petrophysicist, SPDC

Preye Iditumi, Geologist, SPDC

Ukeko Onoriode, Petrophysicist, SPDC

Omodu Meshach, SPDC

Prince Nwogbo, Geophysicist, SPDC

Khalid Amin Khan and Gulraiz Akhter

Special thanks to my friends, Ibikunle Taylor and Joseph Uhrie, for their support.

With deep love and heart full of thanks to my uncle, Samuel Enwenede, cousin, Akpoyibo Ovie, and Popo Enwenede.

My appreciation goes to the following companies, professional bodies and institutions for their materials, data, illustrations and photographs. I was privileged to study and use for this research

- AAPG
- arCIS SEISMIC SOLUTION
- British Petroleum
- Choice Geophysical
- CGGVERITAS
- CSEG
- Earthquake.usgs.gov
- ExxonMobil International
- FairfieldNodal
- FUGRO
- GeoApexTec Inc.

- Geo-Energy Services
- Geophysical Data System
- Geotrace
- Halliburton
- Institute of Petroleum Geosciences Studies, University of Port Harcourt, Nigeria
- International Human Resource Development Corporation
- Journal of the Geophysics Society
- Kansas Geophysical Survey
- Land Ocean Energy Services Co., Ltd.
- Nigeria Association of Petroleum Explorationists (NAPE)
- Oil & Gas Journal
- Paradigm
- Petroleum Seismology Network
- PGS Geophysical
- Pemex-SeaBed Geophysical
- Quality Seismic Services
- Reservoir Geophysics and Geophysics Group, LinKedIn
- Seismic Atlas of the Southeast Asian Basin
- Seismic Processing Network, LinKedIn
- Schlumberger
- Shell International
- Society of Exploration Geophysicists (SEG)
- The Oil and Gas Exploration Network, LinKedIn
- Oil and Gas UK

- Universidade Fernando Pessoa, Porto, Portugal
- WesternGeCo
- World Oil Online Magazine

With special thanks to the below for their help in editing and reviewing the manuscript:

Prof. John O. Etu-Efeotor
Director, Centre for Petroleum Geosciences
Institute of Petroleum Studies
University of Port Harcourt
Rivers State, Nigeria

Dr. James Edet
Exploration Geoscientist
Total E&P Nigeria limited

Dr. Rocco Detomo Jr
Reservoir Geophysicist
Shell International Exploration and Production Inc.
Houston, Texas

Dr. Omu Ugborugbo
Geophysicist
Shell Petroleum Development Company of Nigeria Limited

Above all, I thank God, my Lord and savior Jesus Christ and the sweet Holy Spirit for granting me the wisdom, grace and intelligence, all through this book research.

Foreword

From the primitive forked wood search by blindfolded natives to the modern geophysical and geological methods, the art of searching for hydrocarbons has come a long way. This book 'Seismic Data Analysis Techniques in Hydrocarbon Exploration' is a synthesis of concepts used in the industry to analyse seismic data. It is quite a unique book on seismic data analysis because it is highly practical and focuses on the industry. It is an extensive industry research on seismic data by a young geophysicist who has been exposed to practices in the oil and gas industry. This book explains key concepts and principles in a simplified way that would enhance the understanding of graduate geoscientists and practitioners. Most of the information contained in this book are what a graduate geoscientist requires to work successfully in the oil and gas industry.

I am pleased to recommend this well-researched book to all graduate geoscientists, teachers and practitioners in the field of oil and gas exploration. The organization of the chapters is sequential. Part 1 deals with basic geology and seismic data acquisition in a simplified way. Part 2 discusses the detailed practical steps use in processing seismic data, while the final part deals with seismic data interpretation. This book, therefore, guides the reader from the rudimentary to the expert stages of seismic data analysis techniques in hydrocarbon exploration. This book is easy to read and understood by all classes of hydrocarbon searchers.

I strongly recommend it to all lovers of geosciences.

Professor John O. Etu-Efeotor
Director, Centre for Petroleum Geosciences
Institute of Petroleum Studies
University of Port Harcourt, Rivers State,
Nigeria

Introduction

Seismic data have become an important tool for development of oil and gas field as well as for monitoring oil and/or gas production and not just as an exploration tool. Because of the importance of seismic data to the oil and gas industry, graduate geoscientists need to have a clear understanding of seismic data.

Good understanding of the increased role of seismic technology in oil and gas exploration will enhance the employability of fresh geosciences graduates and make them function more effectively and integrate faster when working within an integrated geosciences team.

The aim of this practical handbook is to explain the fundamental concepts and the detailed practical steps that are needed to understand and interpret seismic data that would enhance the employability of graduate geoscientists into the oil and gas industry.

This book 'Seismic Data analysis Techniques in Hydrocarbon Exploration' was developed from an idea conceived while working as a postgraduate researcher with Shell, Nigeria. I wanted to understand seismic sections and I was convinced that there are fundamental concepts, in which, if I have a clear thought on will aid my understanding of seismic data and how to interpret seismic data. With this idea, I engaged in an extensive industry researched for about 3 years and seek the guidance of highly experience geosciences professionals to write this practical book.

The practical handbook is divided into three chapters: Part 1 explains how sedimentary basin is formed, oil and gas formation, oil and gas traps, application of seismic technology to locate oil and gas traps, how seismic data are acquired, the difference between 2D and 3D seismic data, 4D seismic streamer surveys and OBN survey.

Part 2 explains the detailed practical steps used to extract the geological section of the earth from the acquired seismic field data. Part 3 explains seismic interpretation skills and key fundamental concepts that will aid the readers' understanding on how to predict hydrocarbon directly from seismic data.

The concepts of the book are illustrated using seismic sections, well data and photographs to enhance the readers' understanding of seismic data.

It is my sincere hope that this practical book will meet the needs of graduate geoscientists worldwide who are interested in working in the oil and gas industry, as well as refreshes the mind of industry professionals.

THINGS TO LEARN WHEN YOU READ 'THE PRACTICAL HANDBOOK'

- Graduate geoscientists will get to understand how seismic data are acquired and key fundamental principles and concepts that are applied in seismic exploration.

- Graduate geoscientists will get to understand the detailed step-by-step processing techniques used to convert the acquired seismic data into the geologic section of the earth.
- Graduate geoscientists will get to understand true amplitude (AVO) processing techniques.
- Graduate geoscientists will get to understand geologic structures on seismic sections and why not to drill a buried-focus anticline.
- Graduate geoscientists will get to learn important seismic interpretation skills.
- Graduate geoscientists will learn in practical terms how to interpret fault and do structural and horizon interpretation.
- Graduate geoscientists will learn how to generate geological map and understand time-to-depth conversion technique.

- Graduate geoscientists will get to understand DHI such as 'bright spot', 'flat spot', 'polarity reversal' and 'dim spot', and how to interpret them on seismic section.
- Reservoir engineers will get to understand and learn how wrong seismic velocity could affect wrong estimation of hydrocarbon reserve (STOIP) in the subsurface.
- Geoscientist/petroleum engineers will get to understand 4D (3D-time lapse) seismic survey, its limitations and the use of OBN technology for reservoir monitoring and management as well as to optimize static and dynamic models of complex reservoirs.

And much more about seismic data!

BASIC SEDIMENTOLOGY AND SEISMIC DATA ACQUISITION

CHAPTER
1

Sedimentation and Oil/Gas Formation

GEOLOGIC TIMESCALE OF THE EARTH

Figure 1.1 can be used to represent the geological period of the earth over time. The geology time of the earth's past has been organized into various units according to events which took place in each period. The major divisions of the geologic timescale are called Eons. Eons are divided into eras, which are in turn divided into periods, epochs and ages. The eras of the geologic timescale from oldest to youngest are Precambrian, Paleozoic, Mesozoic and Cenozoic. Note that an era is a subdivision of geologic time that is longer than a period but shorter than eon.

Facts from radiometric dating indicate that the earth is about 4.5 billion years old. The era between this date and the time that larger life forms first appeared in the fossil record is referred to as the Precambrian.

There was life during the Precambrian.

Approximately 600 million years ago during the PALEOZOIC era, life forms such as shellfish started to develop; about 250 million years later life was found on land during the Carboniferous and Permian period.

The MESOZOIC era includes the Triassic, the Jurassic and the Cretaceous. The appearance of dinosaurs marked the beginning of the MESOZOIC era, which is dated as 225 million years ago.

Eon	Era	Period		Epoch	m.y.
Phanerozoic	Cenozoic	Quaternary		Holocene	
				Pleistocene	1.5
		Neogene		Pliocene	
				Miocene	23
		Paleogene		Oligocene	
				Eocene	
				Paleocene	65
	Mesozoic	Cretaceous			
		Jurassic			
		Triassic			250
	Paleozoic	Permian			
		Carboniferous	Pennsylvanian		
			Mississippian		
		Devonian			
		Silurian			
		Ordovician			
		Cambrian			540
Precambrian		Proterozoic			2500
		Archean			3800
		Hadean			4600

FIGURE 1.1 Major division of geological time of the earth. *Source: Explore Montana Geology.*

Mammals are characteristic of the youngest era, the CENOZOIC, which started some 65 million years ago when the first grazing and Carnivorous mammals appeared in a period known as the Tertiary. The Tertiary periods are Pliocene, Miocene, Oligocene, Eocene and Paleocene. These five Tertiary periods make up the CENOZOIC era.

BASIN FORMATION

Sedimentary basins are formed over hundreds of millions of years by the combined action of deposition of eroded material and precipitation of chemicals and organic debris within water environment (Figure 1.2).

FIGURE 1.2 Conceptualized basin formation.

Over time continuing sedimentation occurs in the water environment and the additional weight caused subsidence. Organic matter and different materials deposited at different times, over thousands of years, will produce regular 'layering' of strata in the basin. Volcanic action, or the movement of the earth's crust, causes faults to appear in the basin. The fault formed is conceptualized in Figure 1.3.

FIGURE 1.3   Conceptualized fault formation.

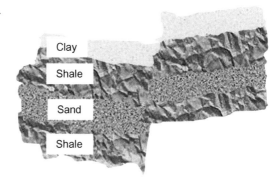

Erosion of the elevated land areas and additional subsidence eventually forms another area of low-lying land (Figure 1.4).

FIGURE 1.4   Conceptualized the surface where erosion has occurred.

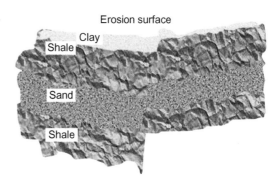

This low-lying land then fills with water forming another water environment. Then, additional sedimentation takes place, causing an 'unconformity' in the underlying strata.

An unconformity is a surface of non-deposition that separates younger strata from older rocks and indicates an interruption in the geological record. Layering below and above the unconformity may or may not be parallel to each other (Figure 1.5).

Finally, land mass movement causes folding and distortion. This gives rise to thick, complex structures and resulted in the formation of sedimentary basin (Figure 1.6).

Note: A stratum is an individual layer (sand or shale) of sediments that is marked at its top and bottom by definite change in lithology or physical break in sedimentation. The plural is strata.

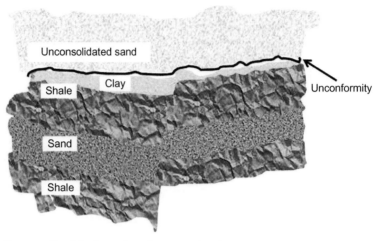

FIGURE 1.5   Conceptualized the formation of an unconformity.

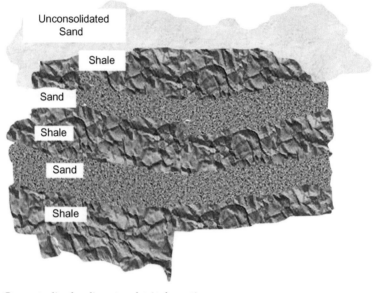

FIGURE 1.6   Conceptualized sedimentary basin formation.

ROCK TYPES

The earth is composed of three basic layers: the core, the mantle and the crust. The crust is the layer that is of the most importance in petroleum geology. The earth's crust is composed of three basic rock types:

- Igneous rock
- Metamorphic rock
- Sedimentary rock

Igneous Rock

Igneous rocks were formed by the cooling and subsequent solidification of molten mass of rock material known as magma. The rate of cooling of the material has a major influence on whether the material is coarse grained or fine grained. One can distinguish between coarse-grained rocks and fine-grained volcanic rocks.

Fine-grained volcanic rocks were formed through rapid cooling and solidification of magma, while coarse-grained rock becomes coarse grained, as the timescale for solidification was much longer.

Metamorphic Rock

Metamorphic rocks are rocks formed from pre-existing rocks, the mineralogy, chemistry and texture of which have been changed by heat and pressure deep within the earth's crust. Shale metamorphosed to slate in a low-temperature environment. At higher temperatures and pressures, shale and siltstone completely re-crystallize forming schist and gneiss rocks. Thus, examples of metamorphic rocks are slates, marble, schist and gneiss, as well as quartzite.

Sedimentary Rock

The sedimentary rocks are formed from either the eroded fragments of upland regions containing older rocks or chemical precipitation. These eroded products are transported by water, wind or ice to the sea, lakes or lowland areas where they settle out and accumulate to form clastic (fragmented) sediments.

Sediments lithify by both compaction, as the grains are squeezed together into denser mass than the original, and cementation, as minerals precipitate around the grains after deposition and bind the particles together. Sediments are compacted and cemented after burial under additional layers of sediment.

As a result, sandstone forms by lithification of sand particles and limestone by the lithification of shells and other particles of calcium carbonate. Evaporation of seawater in contained areas can give rise to the deposition of salts on the sea floor. Anhydrites, calcium carbonate and variety of salts created under such circumstances are referred to as evaporates.

During this processes, trapped organic material gives rise to the oil and gas we are trying to find using seismic technique. These later harden to form the sedimentary rocks such as conglomerates, sandstone, siltstone and mudstone or shale.

Two other types of rocks can be deposited in a marine environment. Abundant calcium carbonate-secreting organisms, including certain algae, corals and animal shells, can be the origin of the formation of limestone, which consists mainly of calcite. Later, chemical alteration may change calcite into dolomite:

$$2CaCO_3 + Mg^{2+} = CaMg(CO_3)_2 + Ca^{2+}$$

Calcite + magnesium ions = Dolomite + calcium ions

Note: Sedimentary rocks are the most important type of rock to the petroleum industry because most oil and gas accumulations occur in them. All petroleum source rocks are sedimentary rock.

OIL AND GAS FORMATION

Sedimentary basins exist around the world at the edges of continental shelves. Hydrocarbons originate from the source rock in a sedimentary basin. Hydrocarbon is formed by organic evolution. Oil and gas are generated when large quantities of organic (plants and animals) debris are continuously buried in deltaic, lake and ocean environment.

Note: A delta is a body of sediment deposited at the mouth of a river or stream where it enters an ocean or lake.

These organic debris are buried rapidly in the subsiding sedimentary basin. With continuing sedimentation and increasing overburden pressure due to increasing weight, sediments containing the organic debris move deeper into the earth. Organic debris normally decays in the presence of oxygen, but with increasing depth, the sediment protects the organic matter by creating an oxygen-free environment. This allows the organic matter to accumulate rather than be destroyed by bacteria. Temperature increases with depth within the earth, so sediments and the organic debris they contain heat up as they become buried under younger sediments. As the heat and pressure continue to build up over millions of years, chemical action occurs, converting the organic debris into kerogen (Figure 1.7).

FIGURE 1.7   Kerogen formation.

Kerogen is of great geological importance because it is the substance that generates oil, gas and coal. Which of these products is being formed is dependent on what organic debris are decomposing and the temperature and pressure it has been subjected to over thousands of years. At temperature below 150 F, the process of hydrocarbon generation seems to be very slow, and at temperature range between 225 and 350 F, the process attains its maximum. As the temperature increases further the source rock gets hotter, chains of hydrogen and carbon atoms eventually break away and form heavy oil. This is conceptualized in Figure 1.8.

FIGURE 1.8   Conceptualized heavy oil formation.

If the temperature continues to rise, the heavy hydrocarbon is converted to lighter light oil or gas. Gas may also be formed directly from the decomposition of kerogen from the woody part of plants (Figure 1.9).

FIGURE 1.9   Depict oil and gas reservoir.

The oil and gas produced by these processes may be in any combination and are always mixed with water. The hydrocarbon produced within the source rocks is then migrated into the pores of the permeable rocks (sandstone). Being lighter than the water in the permeable rock, these hydrocarbons migrate, that is, they move up through the rock until prevented from doing so by an impermeable rock (shale), where they coalesce into larger volumes. A pocket of oil or gas is now formed. Depending on the size of the reservoir, the hydrocarbon formed could become a profitable oil and gas field.

Note that the depth range over which oil generation occurs is known as the 'oil window' and it is usually different for most sedimentary basins.

Note also that commercial deposits of hydrocarbon are pool, field or province.

A pool is the simplest unit of commercial occurrence of hydrocarbon in the subsurface. It is a collection of oil/gas in the same reservoir under the same pressure and temperature system and within a single trap.

Oil and gas field is a collection of hydrocarbon pools that are related to the same geologic feature, structurally or stratigraphically.

Geological regions in which numerous oil and gas fields are located constitute a province.

A reservoir is a subsurface volume of porous and permeable rock that has both storage capacity and the ability to allow fluids to flow through it.

Finally, at temperatures above 500 F, the kerogen becomes CARBONISED, and hydrocarbons are no longer formed. This is conceptualized in Figure 1.10.

FIGURE 1.10   Conceptualized carbonization of kerogen.

OIL AND GAS TRAPS

The movement of the earth's crust causes many different structures to be formed in a sedimentary basin. As such oil and gas are found in different oil and gas traps. The various traps are discussed below.

Type of Traps

Structural Trap

A structural trap is a type of geologic trap that forms as a result of changes in the structure of the subsurface, due to tectonic, diapiric, gravitational and compactional processes. These changes block the upwards migration of hydrocarbons and can lead to the formation of a hydrocarbon reservoir.

The three basic forms of structural traps are the fault trap, the anticline trap and the salt dome trap.

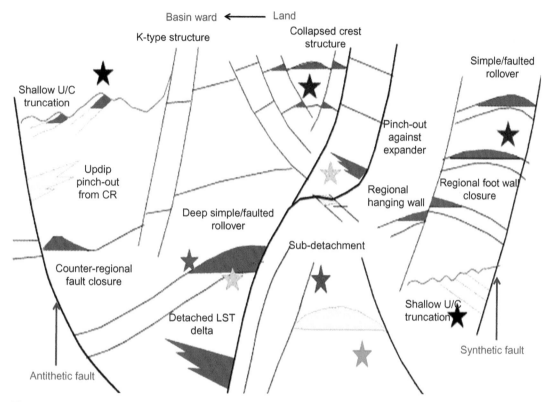

FIGURE 1.11   Different types of fault traps found in the Niger Delta, Basin, Nigeria. *Source: Shell E&P.*

SYNTHETIC FAULT

The fault that is dipping in the direction of the regional sedimentary dip (basinward) is called synthetic fault. This is shown in Figure 1.11.

ANTITHETIC FAULT

The fault that is dipping against the direction of the regional sedimentary dip (counter basinward) is called antithetic fault or counter-regional fault. This is shown in Figure 1.11.

HORST

A horst is an up-thrown block lying between two down-thrown fault blocks.

GRABEN

A graben is a down-thrown block lying between two up-thrown fault blocks. An example of a graben can be seen at the centre of a collapsed crest structure (Figure 1.12).

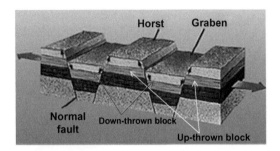

FIGURE 1.12   Horst and Graben. *Source: USGS Earthquake Glossary.*

FAULT BLOCK

A fault block is defined as a subsurface area bounded on either side by faults (Figure 1.11). A fault block could be up thrown (foot wall) or down thrown (hanging wall).

HANGING WALL

The particular stratum (layer) that is resting on the dip direction of the fault plane is the hanging wall (Figure 1.11).

FOOT WALL

The particular stratum that is resting opposite the direction of dip of the fault plane is the foot wall (Figure 1.11).

COLLAPSED CREST STRUCTURE

A collapsed crest structure (this is shown in Figure 1.11) describes a structural play type defined by a series of synthetic and antithetic faults interplaying about a (central) 'point' called the crest. The crestal area may form a bowl which is called a graben.

GROWTH FAULT

Growth faults result from compaction disequilibrium and are found most often in deltas with prolific net sediment accumulation and high shale/sand ratios.

Growth faults are smooth and concave basinward. The name 'growth fault' derives from the fact that after the formation the fault remains active, thereby allowing faster sedimentation in the down-thrown block relative to the up-thrown block.

Structural traps like roll-over anticlines, folds, horst and grabens, which resulted from growth fault, facilitate the entrapment of hydrocarbon. These structural traps are the real trap and not the growth fault itself (Figure 1.13).

Note: In the Niger Delta, Nigeria, the roll-over anticlines form the traps rather than the growth fault that gives rise to them. Almost all the oil fields discovered so far in the Niger Delta complex are associated with roll-over anticlines.

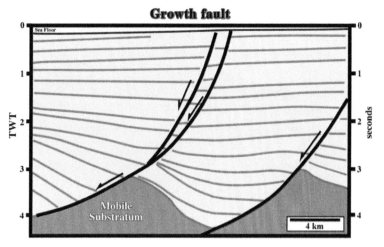

FIGURE 1.13   Growth fault. *Source: Basic Principle in Tectonics. Universidade Fernando Pessoa Porto, Portugal, www.homepage.ufp.pt.*

NORMAL FAULT

When the hanging wall moves down in the direction of dip of the fault plane, it is termed a normal fault or gravity fault. Normal fault is a genetic part of the process of basin formation because they are syn-depositional and dip towards the subsiding basin. Hydrocarbons can be trapped against normal fault, where both a fault plane and an impermeable layer on top of the reservoir form the seal (Figure 1.14).

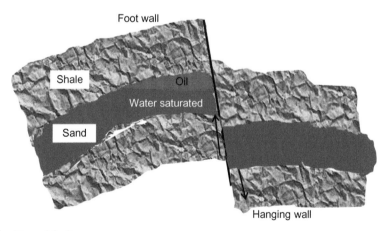

FIGURE 1.14   Normal fault.

REVERSE FAULT

When the hanging wall moves up along the dip direction relative to the foot wall, it is termed a reverse fault (Figure 1.15).

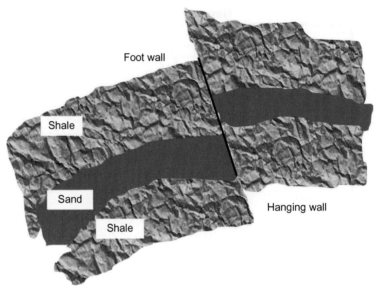

FIGURE 1.15   Reverse fault.

LISTRIC GROWTH FAULT

Listric faults are curved normal faults in which the fault surface is concave upwards; its dip decreases with depth.

The formation of a roll-over anticline will occur when a listric fault collapsed (Figure 1.16).

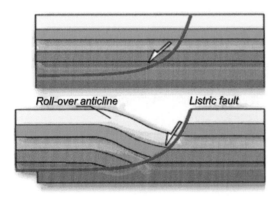

FIGURE 1.16   Listric growth fault. *Source: Structural Geology of Sedimentary Basin by Philip Hutson et al.*

Anticline Trap

An anticline is a structural trap formed by the folding of rock strata into an arch-like shape. The rock layers in an anticlinal trap were originally laid down horizontally and then earth movement caused it to fold into an arch-like shape called an anticline. This type of structure is visible on a seismic section, but, as seismic sections are normally 'time sections', the actual structure may be obscured by velocity changes above the anticline. In an anticline, the structural contours form closed loops (Figure 1.17).

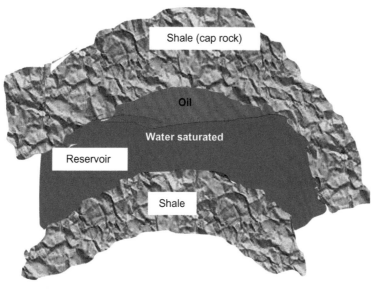

FIGURE 1.17   Anticlinal shape.

The reservoir rock layer in an anticline must be overlain by fine-grained cap rock which seals the top and sides. The closure of the anticline trap is the height of the crest above the lowest structural contour that is closed. It is therefore the part or depth of the anticlinal structure that can hold or contain hydrocarbon. If the contour below does not close any hydrocarbon below, it will move into the next structure by spilling. The structure is usually filled with oil or gas.

If there is more than enough gas to saturate the oil, the excess gas will lie on top of the oil (Figure 1.18). This gives rise to the classic gas–oil–water configuration found in most anticline traps and other hydrocarbon traps.

FIGURE 1.18   Excess gas on top of the oil.

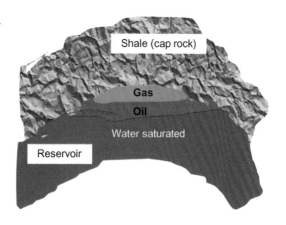

Salt Dome

Salt dome is a trap created by intrusion of stratified rock layers from below by ductile non-porous salt. Hydrocarbon can be found at the flanks of a salt dome and can be found in the deformed layers above a salt dome (Figure 1.19).

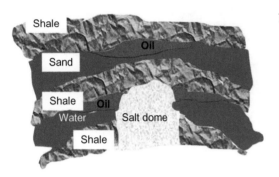

FIGURE 1.19   Salt dome.

Stratigraphic Trap

Stratigraphic traps are formed when there are changes in lithology, nature of the strata or depositional pattern. They prevent continued migration of hydrocarbons within reservoir beds.

A sand body originating from river sand may shale out laterally into the area where clays had been deposited in swamps. The sand body may be a good reservoir, while the shale is a good seal (cap rock). The stratigraphic trap (Figure 1.20) is the most difficult to find on a seismic section.

FIGURE 1.20   Conceptualized a stratigraphic trap.

So far, we have explored how sedimentary basin is formed and how oil and gas are formed and trapped in the subsurface. The exploration geophysicists need to be able to look into the various layers of the earth to determine where oil and gas is trapped. In other to do this the geophysicists apply seismic reflection technology to locate oil and gas traps in the subsurface.

2

Understanding Seismic Wave Propagation

The oil industry uses seismic reflection measurements to gain knowledge about the geological structures in the subsurface in other to locate oil and or gas reservoirs. This method works by sending an acoustic or pressure wave into the earth, which get reflected back when it meets a geological boundary where there is contrast in rock properties. An example of geological boundary is a boundary between sand and shale layers. This wave is then reflected back to the surface after a few seconds and it is then recorded by a receiver. The time that it takes for the wave to return to the surface tells us how far the geological boundary is located (Figure 2.1).

Seismic survey is based on the theory of elasticity and therefore tries to deduce elastic properties of rocks, such as acoustic impedance, Poisson's ratio and V_p/V_s ratio, by measuring their response to elastic disturbances called seismic (elastic) waves.

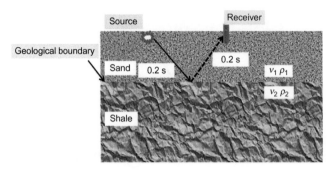

FIGURE 2.1   Conceptualized reflected seismic ray.

TYPES OF SEISMIC WAVES

Two main types of waves are considered in seismic exploration: body wave and surface wave. Surface waves move along the surface of the earth and they are of lower frequency than body waves.

Body Waves

Body waves propagate through the earth's inner layers, and they are further classified as P waves and S waves. These waves are of higher frequency than surface waves.

P Waves

P waves are formed when energy is applied exactly at right angles to a medium. Particle motion under the influence of the wave is then in the direction of propagation of the wave. As a result of the particle motion, the rock particles are alternatively compressed and rarefracted or pulled apart as the waves propagate. P waves are also known as compressional waves, because of the pushing and pulling they do. They are the fastest kind of seismic waves.

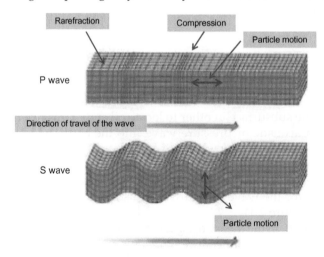

FIGURE 2.2   Conceptualized P- and S-wave propagation. *Source: USGS.*

In seismic processing, one generally assumes that we are dealing with P waves (Figure 2.2) only, travelling through isotropic and homogeneous material.

If the elastic moduli (K) and the bulk density (ρ) of the rocks are known, P-wave velocity can be calculated using the following equation:

$$V_P = \sqrt{\frac{K + 4/3\mu}{\rho}}$$

CHARACTERISTICS OF P WAVES

- It is a sound wave or pressure wave.
- It is always the first arriving seismic wave.
- Its velocity is the fastest of all seismic waves.
- Its particle motion is in the direction of propagation of the wave.
- It can propagate in solids or fluids (oil, gas and water).

S Waves

S wave, also called shear waves, is the second type of body wave. S wave is formed when energy is applied in a direction parallel to the surface of a medium. S wave does not propagate through fluids. Shear wave exploration cannot be done with conventional marine seismic acquisition technology. Ocean bottom nodes (OBN), in which receivers are mounted on the seabed (Figure 2.3), can be used to measure S waves.

Remotely operated vehicle (ROV)   Node on the seabed

FIGURE 2.3   Trilobit 4C ocean bottom nodes deployed on the seabed using ROV (remotely operated vehicle) system. *Source: WorldOil Online Magazine, vol. 232(9), September 2011.*

S-wave velocity can be calculated using the following equation:

$$V_s = \sqrt{\frac{\mu}{\rho}}$$

P-wave velocity is always faster than S-wave velocity. The ratio of P-wave velocity to S-wave velocity is greater than or equal to the square root of 2

$$\frac{V_{\mathrm{P}}}{V_{\mathrm{s}}} \geq \sqrt{2}$$

CHARACTERISTICS OF S WAVES

- The particle motion under the influence of the wave is in the direction of the energy applied and is then perpendicular to the propagation of the wave.
- S waves can only propagate in solid.
- There is no compression and expansion as a result of the particle motion, and thus no pressure disturbance.
- One can detect S waves through the measurement of the particle motion.

WAVE PARAMETERS

Figure 2.4 illustrates some important wave parameters that are applicable in seismic exploration.

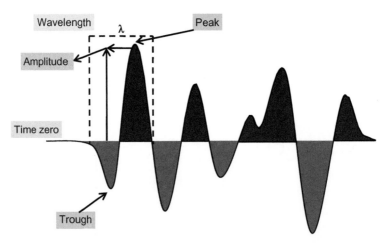

FIGURE 2.4   Sinusoidal waveform (seismic trace).

A sinusoidal waveform is described by the mathematical expression below:

$$\varphi\,(t, x) = A \cos\,(2\pi ft + \phi)$$

The above equation described a seismic trace.
The parameters in the above equation are defined below:

- The amplitude is the maximum displacement from time zero.
- Period (t) is the time, in seconds, required to complete 1 waveform.

- Frequency (f) is the number of times a waveform repeats per second. It is the reciprocal of period. Frequency is measured in Hertz (Hz), where 1 Hz equals 1 waveform/s.

$$f = \frac{1}{T}$$

For example, a waveform with duration of 30 ms has a frequency of 33.3 Hz. Note that 1 s equal to 1000 ms:

- A wavelength (λ) is the distance over which one waveform is completed. It is measured in metre or feet:

$$\text{wavelength } (\lambda) = \frac{\text{velocity } (v)}{\text{frequency } (f)}$$

- Wave number is the reciprocal of wave length:
  wave number = 1/wavelength.
- Phase indicates the initial position of the waveform. Phase is an angular quantity that is measured in radians or degrees.

There is a very important relationship between frequency (f), wavelength (λ) and velocity (v) for all waves. This is given by

$$v = f\lambda$$

where v is seismic velocity, f is frequency of the propagating wave and λ is the wavelength.

SEISMIC REFLECTION

Law of Reflection

The law of reflection states that the angle of incidence equals the angle of reflection. It also states that the incident ray, the reflected ray and the normal at the point of incident all lie in the same plane (Figure 2.5).

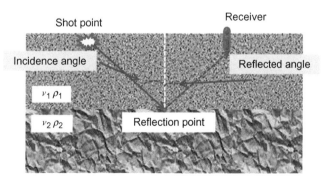

FIGURE 2.5   Conceptualized law of reflection.

For a horizontal reflector, the reflection point is half way from source point to receiver. As a rule of thumb in the industry, the maximum offset (the offset is the distance between the source and the receiver) is equal to the depth of the reflector. For straight rays the angle of incidence is about 28°. But rays in the real earth are not straight. The angle of incidence for real earth is approximately 35° (Figure 2.6).

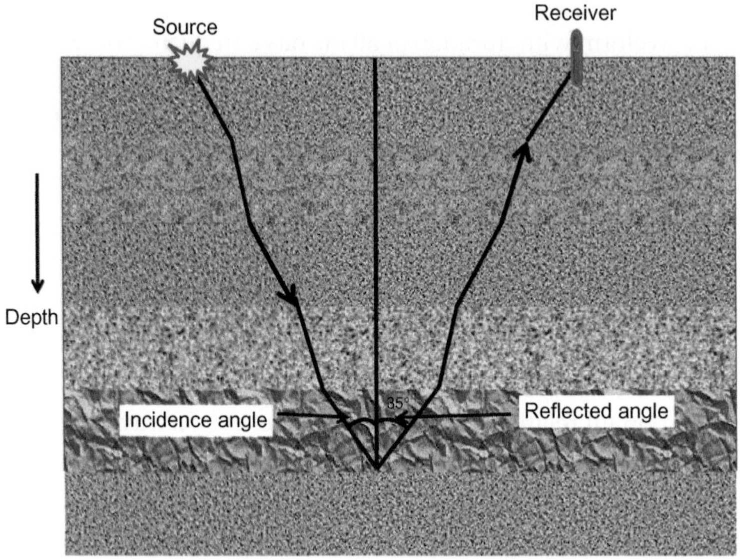

FIGURE 2.6 Conceptualized curve ray path in the earth.

For a dipping reflector, the reflection point does not lie at the midpoint between the source and the receiver but up-dip of the source receiver pair (Figure 2.7).

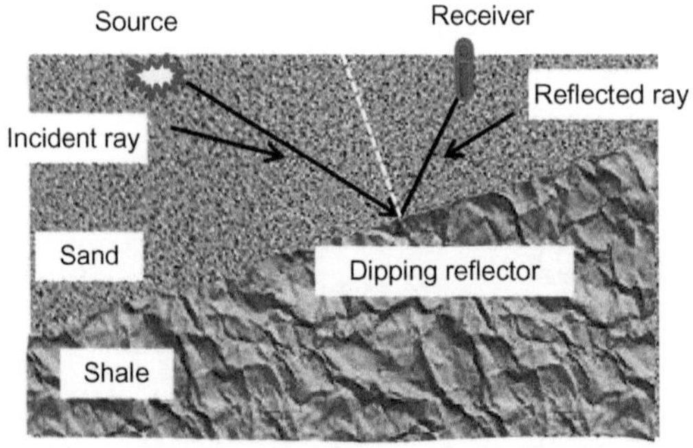

FIGURE 2.7 Reflection point on a dipping reflector.

Note that the incident ray generates not only the reflected ray but also the refracted ray at an angle given by Snell's law.

Law of Refraction

The law of refraction is known as Snell's law. It states that the ratio of the sine of the angle of incident to the sine of the angle of refraction is a constant for a given pair of strata

$$\frac{\sin \theta_i}{\sin \theta_r} = n$$

where n is a constant, θ_i is the angle of incidence and θ_r is the refracted angle.

FIGURE 2.8   Conceptualized Snell's law.

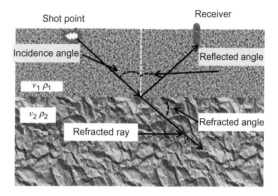

CRITICAL ANGLE

As the angle of incidence increases in Figure 2.8, a point is finally reached where the refracted ray does not emerge at the second layer but lie along the interface. This particular angle of incidence at which the angle of refraction is 90° and the refracted ray lies along the interface is known as the critical angle. At and beyond the critical angle, there is no transmitted ray and therefore a very high reflected ray will be recorded (Figure 2.9).

FIGURE 2.9   Conceptualized critical angle.

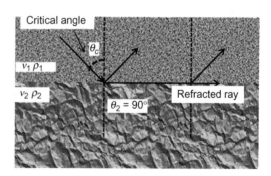

Therefore,

$$\frac{\sin \theta_i}{\sin 90} = \frac{V_{p1}}{V_{p2}}$$

But, sin 90 = 1.
At critical angle,

$$\sin \theta_{critical} = \frac{V_{p1}}{V_{p2}}$$

A critical refracted wave travels along the interface between layers and is refracted back into the upper layer at the critical angle. The waves refracted back into the upper layer are called head waves or first-break refractions because at certain distances from a source, they are the first arriving energy. Recorded first-break refraction is shown in Figure 2.10.

Note that these first-break refractions can give us important information about the shallow velocities on land seismic data.

Note also that seismic data are acquired in such a way that reflections from horizons of interest are in the pre-critical region, even at the farthest offset in the data.

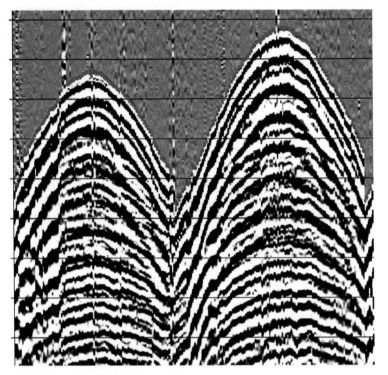

FIGURE 2.10   First-break refraction data.

In reality, part of the seismic energy arriving at an interface is transmitted and refracted, and another part of the energy is reflected at that same interface. Given that there are many reflectors in the subsurface, there will be many paths from source to receiver, each of them with a different travel time. The proportion of energy reflected depends on the material properties of the two bounding layers and on the angle of incidence.

REFLECTION COEFFICIENT

In each layer in Figure 2.11, we have the density, ρ, and velocity, v. The product of velocity and density is a material property of the layers and is known as acoustic impedance. The acoustic impedance is an important property of a rock layer. This is because it determines the reflection response of the earth. In the earth, reflections occur at the interfaces between layers and the reflection amplitudes depend mostly on the difference of acoustic impedance from layer to layer. The term AI unit is used for the unit of acoustic impedance.

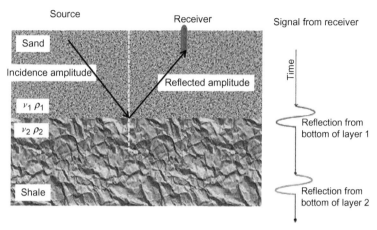

FIGURE 2.11   Reflection coefficient.

Reflected amplitude is the difference between the two impedances of each layer divided by their sum. The ratio of the reflected amplitude to the incidence amplitude is called the reflection coefficient and is given by the difference in acoustic impedance divided by their sum

$$\text{reflection coefficient} = \frac{Z_2 - Z_1}{Z_2 + Z_1}$$

where Z_1 and Z_2 are acoustic impedance of layers 1 and 2, respectively.
     Calculations of reflection coefficients should give results between -1 and $+1$.
     The reflection coefficient can still be expressed as

$$\text{RC} = \frac{V_2\rho_2 - V_1\rho_1}{V_2\rho_2 + V_1\rho_1}$$

The sum in the denominator of the reflection coefficient equation does not have very much effect. It is the difference in the numerator that really dominates the reflection coefficient.
     For example, if we use the reflection coefficient equation and calculated that the reflection coefficient is 0.2, then this implies that a wave with amplitude of 20% of the original amplitude of the wave reaching a reflecting interface is returned towards the surface. As the energy is proportional to the square of the amplitude, 4% of the energy is reflected, and that the remaining energy passes through the interface (Figure 2.12).

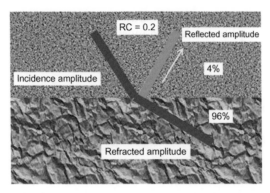

FIGURE 2.12   Reflected amplitude.

Note that reflection coefficient equation is an approximation. It only holds true for rays at right angles to the interface.

Note also that ray path models can be used to determine the actual amplitude of the reflections returned from any layer.

TRANSMISSION COEFFICIENT

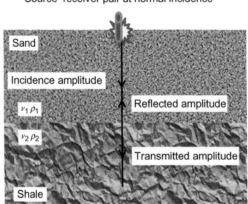

FIGURE 2.13   Transmission coefficient.

Transmission coefficient (Figure 2.13) describes the amplitude of the transmitted wave and is given by the expression

$$\text{transmitted amplitude} = \frac{2Z_1}{Z_2 + Z_1}$$

If the reflection coefficient is R, then a little mathematics gives the transmission coefficient as $1 + R$, so that for a wave going from a softer rock (sand) to harder rock (shale), the transmitted amplitude is greater than the incident amplitude. This is because the transmitted amplitude is measured in a different material from the incidence and reflected amplitudes.

For a wave going downwards the transmission coefficient is $1 + R$ and for a wave going upwards the transmission coefficient is $1 - R$ (Figure 2.14).

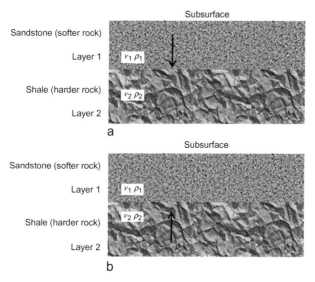

FIGURE 2.14 (a) The arrow shows a down-going wave. Transmission coefficient is $1 + R$. (b) The arrow shows a up-going wave. Transmission coefficient is $1 - R$.

Note: The two-way transmission coefficient is expressed as

$$\text{two-way transmission coefficient} = (1 + R)(1 - R) = 1 - R^2$$

UNDERSTANDING CONVOLUTION

From well log data (velocity and density log), we can construct a lithologic variation such as changes from sand to shale (Figure 2.15).

FIGURE 2.15 Conceptualized stratigraphic layer of the earth.

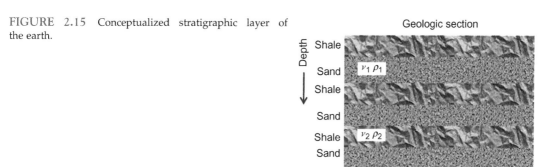

Product of the velocity and density log gives the acoustic impedance (AI) log or profile as shown in Figure 2.16.

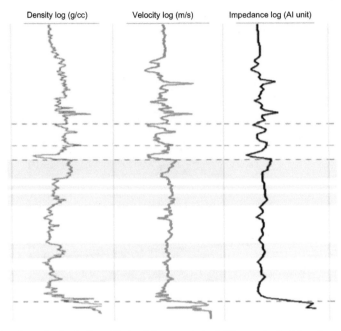

FIGURE 2.16   Density and velocity logs (derived from sonic log). The product of density and velocity logs gives the acoustic impedance log.

Recall that, seismic reflection occurs when there is contrast in acoustic impedance across a layer boundary. From the acoustic impedance log, the geoscientist can calculate the reflection coefficient for each reflecting interface in the subsurface.

These reflection coefficients formed the reflectivity series which is either displayed as time or depth.

The earth reflectivity is a set of impulse response with amplitudes proportional to the reflection coefficient of each reflecting horizon and time of occurrence equal to the two-way reflection time (Figure 2.17).

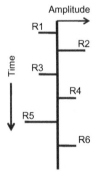

FIGURE 2.17   Earth reflectivity series from each reflecting interface.

In seismic acquisition, the source wavelet that is sent into the earth convolved with the earth's reflectivity series to produce the recorded seismic trace. This process used to derive our seismic trace is called convolution.

Mathematically, convolution is:

$$w(t) \times r(t) = s(t)$$

In other words, convolution is an operation describing the interaction between a unit impulse (source wavelet) and a system (earth reflectivity) to produce an output (seismic trace) (Figure 2.18).

FIGURE 2.18 Conceptualized the convolutional model of the seismic trace.

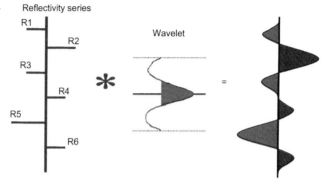

Note that the convolution of any time series (earth reflectivity) with a unit impulse (spike produced from seismic energy) is simply the input time series (seismic trace).

CROSS-CORRELATION

Cross-correlation is a process for measuring the similarity of one time series (seismic trace) to another time series (seismic trace). An example of cross-correlation is shown in Figure 2.19.

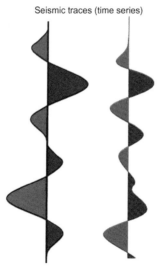

FIGURE 2.19 Conceptualized cross-correlation.

AUTOCORRELATION

Autocorrelation is simply the cross-correlation of one time series with itself.

TIME DOMAIN

Convolution described time series and wavelets in terms of amplitude as a function of time. This is time domain description of seismic trace. That is, in time domain, a seismic trace is described as having certain amplitudes at certain time.

FREQUENCY DOMAIN

In frequency domain, seismic trace is described in terms of amplitudes and phases at certain frequencies.

To describe seismic trace in the frequency domain requires both an amplitude spectrum and a phase spectrum. The amplitude spectrum simply gives amplitude at each frequency. The phase spectrum simply gives the phase at each frequency (Figure 2.20).

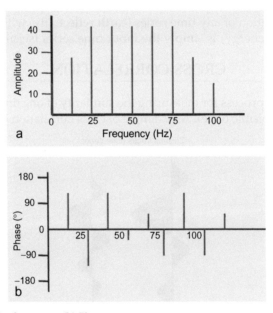

FIGURE 2.20   (a) Amplitude spectra. (b) Phase spectra.

Convolution of our seismic trace in the time domain is equivalent to multiplication of the trace amplitude spectrum with the filter amplitude spectrum and the addition of the trace phase spectrum with the filter phase spectrum in the frequency domain (Figure 2.21).

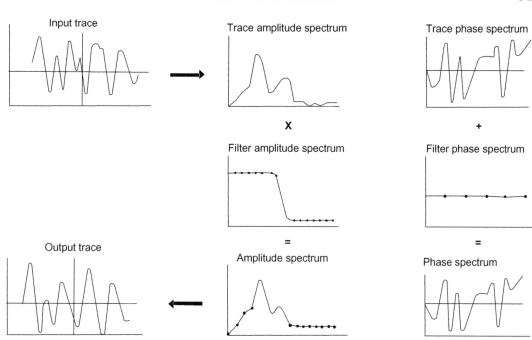

FIGURE 2.21   Conceptualized convolution in the frequency domain.

Correlation in time domain is equivalent to multiplication of the trace amplitude spectrum with the filter amplitude spectrum and subtraction of the trace phase spectrum from the filter phase spectrum in the frequency domain.

Notice that in convolution the phase spectrums are added together, while in correlation the phase spectrums are subtracted one from another. This difference explains why the autocorrelation of any trace is a zero phase.

Note also that autocorrelation of a seismic trace in the time domain is equivalent to multiplying the amplitude spectrum by itself and subtracting the phase spectrum from itself.

FOURIER TRANSFORM

The Fourier transform process allows the geoscientists to transform the seismic trace from the time domain, which is a plot of amplitude versus time, to frequency domain, which is the amplitude and phase spectra.

INVERSE FOURIER TRANSFORM

The geoscientists use inverse Fourier transform to transform the seismic trace from frequency domain to time domain.

The seismic record consists of about 240 traces recorded every 25 m on the surface and thousands of traces are in the total data volume. To transform this quantity of data into the frequency domain, an algorithm has been developed that speeds up the process to allow transformation from time domain to frequency domain and vice versa.

The algorithm to transform from time domain to the frequency domain is called the fast Fourier transform (FFT).

The algorithm to transform from frequency domain to the time domain is the inverse fast Fourier transform (IFFT).

Understanding Seismic Exploration

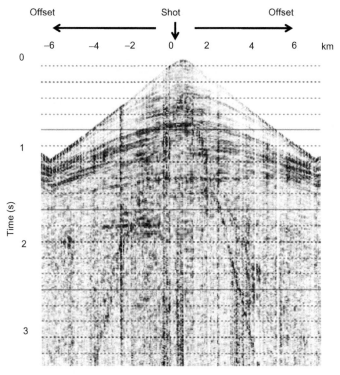

FIGURE 3.1   Seismic field record.

The display above is a raw seismic field data. The horizontal scale is the receivers (geo-phones) numbers which can be translated into metres/kilometres. Two-way travel time is recorded on the vertical axis of the seismic field record in fractions of seconds. Sometimes, it is more convenient to express time as milliseconds. The two-way time is the time required for the seismic energy to radiate from the source at the surface into the sub-surface until it reaches a reflector and reflected back to the surface, where it is recorded by the receiver.

Note that a reflector is a boundary between rocks with different properties. There may be a change of lithology or fluid fill from rock layer to layer. These property changes cause some sound waves to be reflected towards the surface. And there are many reflectors on recorded seismic data.

The receiver number nearest to the first seismic shot in Figure 3.1 is on the right and left (say −2 and 2) in the middle. The receiver number furthest away from the shot is on the right and left (say −6 and 6).

Notice the hyperbolic shape of the reflections in Figure 3.1. This is because near the first shot the energy travels almost straight down and up – very little lateral distance. For receivers far from the shot (say 6 km), the energy has not only a vertical component but also a horizontal component. Based on the hyperbolic shape of the reflections, the geophysicists can calculate the average velocity along the ray paths.

The data recorded by the receiver far from the shot are known as the far-offset data. While that recorded near the shot are known as near-offset data. The distance travelled by the reflected seismic energy is longer for the far offset than the near offset and takes more time.

Let's explore in detail the fundamental concepts use to acquire the raw seismic field data in Figure 3.1.

SEISMIC DATA ACQUISITION

Seismic acquisition starts with a clear goal in mind of what the geoscientists want to image in the sub-surface. For example, a roll-over anticline at 10,500 ft. covering 40 km^2 and sands believed to be 200 ft. thick. This will determine the survey parameters such as survey area, the fold, shooting direction, etc.

Thus, the objective of seismic data acquisition is to obtain data that can be related to the structural image of the sub-surface geology for the purpose of locating oil and gas traps.

Before seismic acquisition is carried out, the area or location where the exploration activities will take place is surveyed by a surveyor and the seismic line is accurately marked out by the surveyor. The lines serve as paths for preparing shot holes and laying of cables for recording.

Global Positioning System (GPS) and total stations are used for setting out the lines. The survey lines usually form grids of 500 – 500-m spacing (Figure 3.2).

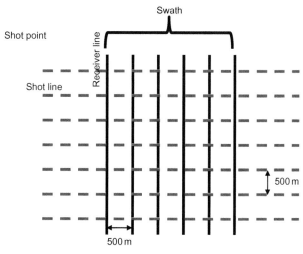

FIGURE 3.2   3D survey lines.

Before going further, let's define some acquisition parameters.

Swath

Swath is the sub-division of seismic survey area into six receiver lines.

Shot Line Interval

The distance from one shot line to the next shot line is called the shot line interval. The shot line is also called cross-line.

Shot Point

Shot points are locations at the surface of the Earth at which a seismic shot is activated.

Salvo

Salvo is the number of fired shot in a shot line. For example, if the number of fired shots in a shot line is 120 and the distance from one shot to the next shot point is 50 m apart; therefore, the shot line length is 120 multiple by 50, which is equal to 6000 m or 6 km.

Shot Point Interval

Shot point interval is the distance between one shot point and the next shot point located on the same receiver line.

Source Density

Source density is the number of shots per surface area. The number of source lines per kilometre and the number of sources per kilometre determine the shot density.

Receiver Line

Receiver line is the line where geophone groups are located at a regular distance. The receiver line is called in-line.

Receiver Interval

Receiver interval is the distance between two receiver groups located on the same receiver line.

Receiver Line Interval

Receiver line interval is the distance between two consecutive receiver lines.

Receiver Density

Receiver density is the number of receivers per surface unit. The number of receiver lines per kilometre and the number of receivers per kilometre determine the receiver density.

Azimuth

Azimuth refers to the direction of the line from shot to receiver.

Template

Template represents all active receivers corresponding to one given shot point (Figure 3.3). Shot point can be inside the template or outside.

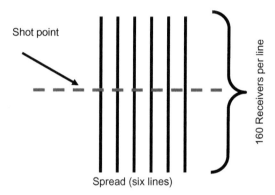

FIGURE 3.3   A template.

In-Line Move-Up

After the completion of the shots per salvo, the template is moved up to the next shot line (salvo). This process is called in-line move-up.

Cross-Line Move-Up

This occurs when the template reaches the edge of the survey area and moves up across the survey to start a new in-line move-up.

Patch

Patch is an acquisition technique where source lines are not parallel to receiver lines.

Geometrical Parameters

Geometrical parameters correspond to offsets, source, and receiver layouts.

Recording Parameters

Recording parameters are related to recording length and sampling rate.

ONSHORE SEISMIC SOURCE

Seismic source is a device that provides acoustic energy for acquisition of seismic data. Examples of seismic source are explosive charge, vibrator, and airgun.

Onshore seismic data are acquired using an explosive energy source, such as dynamite.

Dynamite is used as the seismic source to generate acoustic waves that are sent into the sub-surface. The detonation of an explosive (dynamite) is referred to as the seismic 'shot'. The seismic shots are normally placed below the weathered (low velocity) layer of the earth. This improves the coupling of the seismic source to the sub-surface and avoids problems with the very variable acoustic velocities in the weathering layer.

Shot holes are drilled at shot points and dynamite is placed in the holes. Drill casing and hand auger are used for shallow pattern holes usually 3.5–4 m deep. In swamp terrain the shot holes are usually 45–60 m deep. The explosives (about 2 kg) are placed into the drilled shot holes (shot point). Each shot position is numbered, and the position of each shot (shot point) is accurately mapped. With other onshore seismic sources, such as vibrators and shots from air shooting, the shots occur at the Earth's surface.

The number of shots and their positions are carefully designed to improve the downward-going energy and to attenuate the energy going in other directions.

FIGURE 3.4   Shows drillers drilling shot hole and also place the charge in shot hole.

Note that the size of the explosives used depends on the objective of the survey, in particular, on the depth of the target reflectors. One kilogram of seismic source releases about 5 MJ of energy. Some oil companies use 2 kg of dynamite as their source of seismic energy.

Thus, the energy released in a shot depends on

- The explosive size
- The depth of the shot hole
- The local ground conditions.

The position and elevation of each shot position is carefully recorded, and also the depth of the shot hole is measured. These parameters are required for the correct processing of the acquired seismic data.

Once the shot positions have been marked out, geophone in groups are laid out at regular intervals prior to shooting around or to one side of the shot position typically extending over 3–6 km and are connected by wire to the recording truck and the acquisition work begins. As the shot position advances down the line, different sections of the recording groups (geophone) are made 'live' by the recording instruments. This maintains a similar range of 'offsets' for each shot.

The offset is the distance from the shot to the receiver. At some point, the geophone group must be moved to maintain the same range of offsets in the 'live' section.

Note that changes in the speed (velocity) of sound and density within a particular rock layer cause reflection and refraction of the sound waves produced by a seismic source. Variation of these parameters at an interface between two different rock types causes reflection of some of the seismic energy back towards the surface. It is the record of these reflections against time that produces the seismic section shown in Figure 3.5.

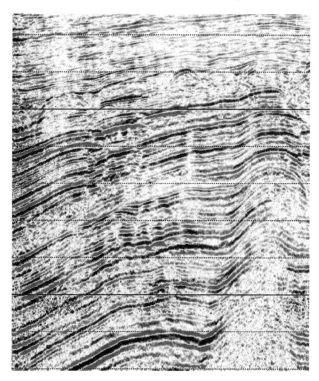

FIGURE 3.5   A seismic section.

ONSHORE SEISMIC RECORDING

Seismic recording is done using a device that detects seismic energy in the form of ground motion or a pressure wave in fluid and transforms it to an electrical impulse.

Onshore seismic data are recorded using a simple electro-magnetic device known as a geophone (Figure 3.6).

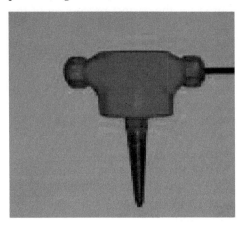

FIGURE 3.6　A geophone. *Source: Geophone.png-commons. wikimedia.org.*

Geophone is used, both onshore and on the seabed during marine seismic acquisition, to detect ground velocity produced by acoustic waves and transform the motion into electrical impulses. The geophone generates current that is proportional to the particle velocity of the earth.

The electrical current flows out of the wires connected to the geophones and goes to an analogue-to-digital converter to form digital data. The data could travel along a cable to the recording unit and be digitized there, or it could come out of the geophone group as analogue data and be digitized at that point to be sent via wire to the recording unit (Figure 3.7).

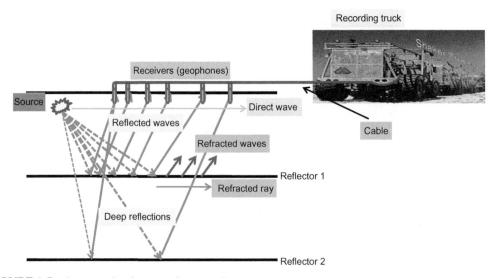

FIGURE 3.7　Conceptualized seismic data recording.

Note that geophones detect motion in only one direction (in the vertical direction).

The amount of energy recorded by one geophone is small. Therefore, several geophones are grouped together in an array around the central receiver position and this is called a channel (Figure 3.8).

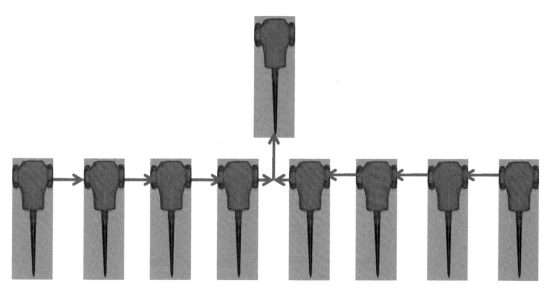

FIGURE 3.8   Geophone string.

The geophone in an array is analogue summed to form the output signal going into the central receiver. This represents one channel going into the recording truck (Figure 3.7).

Simultaneous recording of 500–2000 channels is common during 3D seismic acquisition and 120–240 channels during onshore 2D seismic acquisition.

Note that grouping geophones in an array improves the total signal output from the group and also 'tunes' the geophones so that energy from below is enhanced while those from the side (for instance, ground-roll) is attenuated.

SEISMIC TRACE

Seismic traces are data recorded from one 'shot point' at one receiver position. Seismic traces are recorded as a function of time. As this time represents the time taken for the acoustic energy to travel into the earth, reflect, and then return to the surface, it is correctly called 'two-way time'. It is measured in seconds or milliseconds (Figure 3.9).

FIGURE 3.9   Seismic trace.

The resultant collection of traces recorded from one shot point is generally recorded together and it is referred to as 'seismic field record'.

The horizontal scale of the field record is receiver number which can be translated into ft./miles or metres/km. The vertical scale is a two-way travel time (Figure 3.10).

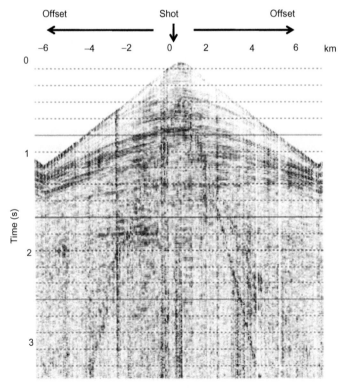

FIGURE 3.10   Seismic field record.

A field record could consist of 240 seismic traces recorded every 25 m on the surface, and so a large amount of data are acquired in a seismic survey.

The display of many traces side by side in their correct spatial positions produces the final 'seismic section' that provides the geologists/geophysicists with a structural picture of the sub-surface (Figure 3.11).

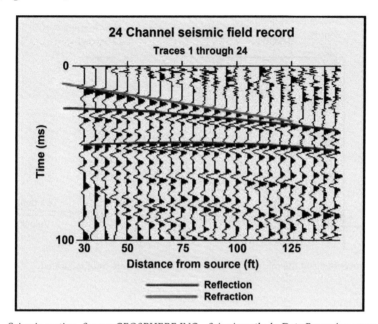

FIGURE 3.11   Seismic section. *Source: GEOSPHERE INC – Seismic methods: Data Processing. www.geosphereinc.com.*

FOLD OF COVERAGE

In 3D seismic processing, the survey area is divided into a grid of 25×25 m called bin (Figure 3.12).

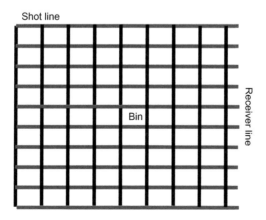

FIGURE 3.12   A bin.

A bin is a square or rectangular area, which contains all traces that correspond to the same common midpoint (CMP). To the geophysicists, seismic traces live at the midpoint of the source–receiver distance (Figure 3.13).

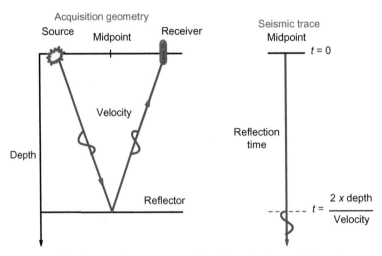

FIGURE 3.13   Conceptualized the midpoint position where the seismic trace is located.

The fold of coverage of a 3D survey is the number of traces that are located within a bin and that will be stacked. Typical values of fold for modern seismic data range from 60 to 240 for 2D seismic data, and 10–120 for 3D seismic data.

CMP Fold

The number of traces that are associated with any given midpoint location and are added during stacking to produce a single trace is termed the CMP fold.

Fold Map

A fold map or multiplicity map shows the total coverage area of a 3D seismic survey.

Nominal Fold

The nominal fold (full fold) of a 3D survey is the fold for the maximum offset. Due to the reflection technique, the fold is not nominal at the edge of 3D survey as shown in Figure 3.14.

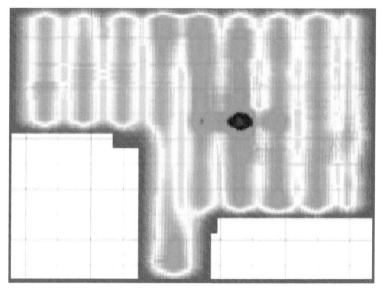

FIGURE 3.14   Fold map. It represents the total coverage area of the seismic survey.

HOW TO CALCULATE FOLD OF COVERAGE

The fold of coverage can be calculated as follows.

In-Line Fold

$$\text{In-line fold} = N_c * R_s/2 * \text{SI}$$

where N_c is the number of recording channels per receiver line, R_s is the receiver interval or spacing, and SI is the shot line interval.

Cross-Line Fold

$$\text{Cross-line fold} = N_s * S_s/2 * \text{RI}$$

where N_s is the number of shot points in a shot line or the number of shots per salvo, S_s is the shot point interval or spacing, and RI is the receiver line interval or spacing

$$\text{Total fold} = \text{In-line fold} * \text{Cross-line fold}$$

Total multiplicity is the percentage of the total fold.
For example, let's calculate the fold from this set-up shown in Figure 3.15.

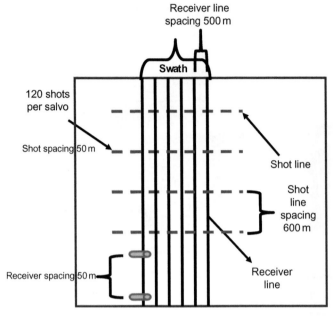

FIGURE 3.15   This set-up is used to describe how fold is calculated.

Shot spacing or interval $(S_s) = 50$ m
Shot line spacing or interval (SI) $= 600$ m
The number of shots in a shot line or shots per salvo $(N_s) = 120$
Receiver spacing or interval $(R_s) = 50$ m
Receiver line spacing or interval (RI) $= 500$ m
Assume the number of receiver channels per receiver line $(N_c) = 240$
Recall that,

$$\text{In-line fold} = N_c * R_s/2 * \text{SI}$$
$$= 240 * 50/2 * 600 = 12,000/1200$$
$$\text{In-line fold} = 10$$

$$\text{Cross-line fold} = N_s * S_s/2 * \text{RI}$$
$$= 120 * 50/2 * 500$$
$$= 6000/1000$$
$$\text{Cross-line fold} = 6$$

Thus,

$$\text{Total fold} = \text{In-line fold} * \text{Cross-line fold}$$
$$= 10 * 6 = 60$$

Therefore,

$$\text{Total multiplicity} = 6000\% \ (60\text{-fold})$$

MARINE SEISMIC SURVEY

Marine seismic surveys are conducted using specially equipped vessels that tow one or more cables containing a series of hydrophones at constant intervals (Figure 3.16).

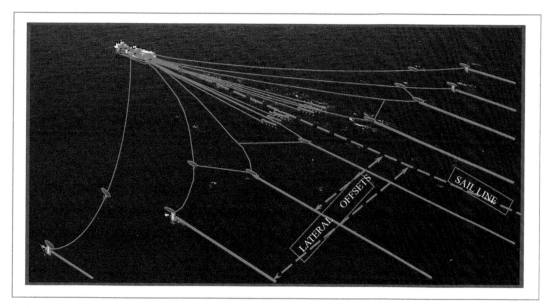

FIGURE 3.16   Multiple sources and receivers behind the seismic vessel.

Three seismic vessels can sometimes be used in marine seismic survey. One of these vessels is used as the main shooting, recording, and processing vessel. The other two vessels towed additional streamers for increased coverage. These two other vessels record seismic signal but are not involved in the shooting or processing of the data. The streamers locations are accurately positioned by the aid of Global Positioning System (GPS).

Note that seismic vessel can easily obtain 10 times more data per day than the best land seismic crew. This is because the hydrophone arrays are towed behind the boat, eliminating the need for continuous laying of new spreads.

OFFSHORE SEISMIC SOURCE

Airguns are used as sources of seismic energy in the acquisition of marine seismic data. This airgun releases a blast of highly compressed air into the surrounding water. Airguns are also used in water-filled pits on land as an energy source during acquisition of vertical seismic profiles.

Following the initial burst of energy, pressure interaction between the air 'bubble' and the water causes the bubble to oscillate as it floats towards the surface. The amplitude and time difference between these bubble pulses depend on the depth of the gun and the size of the main chamber in the gun. The depth of the gun is determined by the bandwidth desired.

The gas bubble produced by an airgun (Figure 3.17) oscillates and generates subsequent pulses that cause source-generated noise.

FIGURE 3.17   An airgun used in marine seismic survey.

A single airgun is not a perfect source. Combination of an array of airguns with different chamber sizes and firing these simultaneously improves effectiveness as well as makes it a better source.

Careful use of multiple airguns can cause destructive interference of bubble pulses and alleviate the bubble effect. A cage, or a steel enclosure surrounding a seismic source, can be used to dissipate energy and reduce the bubble effect.

Depending on the marine acquisition design, each airgun array is composed of three airgun strings. Each string in turn is made up of nine airguns. The airgun array is typically about 23 m long (Figure 3.18).

FIGURE 3.18   An airgun array been deployed into the sea. *Source: oceansjsu.com in Google image.*

MARINE SEISMIC DATA RECORDING

The recording of marine seismic data is done by a streamer. 3D seismic vessel may deploy from 1 to 12 streamers and shooting arrays. These provide the high densities of data required in marine seismic acquisition. The streamers are deployed just beneath the surface of the water and are at a set distance away from the vessel.

All of the recording equipment (hydrophone) is encased in the streamer (Figure 3.19) that is towed behind the seismic vessel.

FIGURE 3.19   A Columbia gad student carries 'bird' for seismic streamer. *Source: The Earth Institute Columbia University and Donna Shillington. Blogs.ei.columbia.edu.*

The hydrophone detects changes in pressure that are created as sound waves from the airguns bounce off geological strata beneath the seafloor. Hydrophones used in marine recording use a pressure-sensitive device to record the incoming energy. Hydrophones are connected together in groups in the streamer and may be placed 6 m apart. Complex electronics within the streamer filters the incoming signal from a whole group of hydrophones and then converts the resultant voltage into a digital format. The numbers corresponding to the values for all of the groups at any one time are interleaved in a process called multiplexing and then sent digitally down just a few wires to the recording instruments. The resultant digital recordings are eventually sent to the processing centre (which may also be on the vessel).

Bird in the streamer and other positioning equipment are used to provide dynamic information about the position of every hydrophone group for every streamer for every shot.

A tail buoy (a floating device) is used to identify the end of a streamer. The front end of the streamer is connected to the seismic vessel by a system of floats and elastic stretch sections. This eliminates any noise reaching the streamer from the seismic vessel. The tail buoys allow the seismic acquisition crew to monitor the location and direction of streamers. The tail buoy attached at the end of the cable may contain its own GPS receiver and radar reflector so that its position can be established (Figure 3.20).

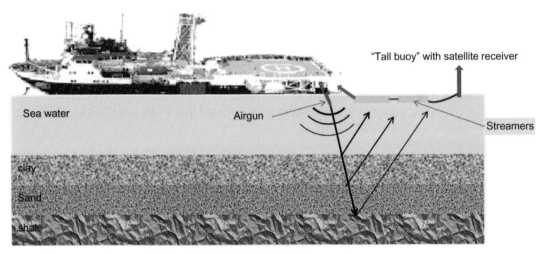

FIGURE 3.20   Conceptualized how the front end of the streamer is connected to the seismic vessel.

TWO-DIMENSIONAL SEISMIC DATA (2D)

In 2D seismic survey, shot and receivers are in the same line (Figure 3.21). And the next line is spaced kilometres away. In other words, 2D seismic data are acquired individually, as opposed to the multiple closely spaced lines acquired together that constitute 3D seismic data.

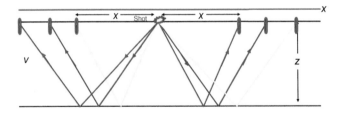

FIGURE   3.21   Conceptualized   2D seismic configuration.

PROBLEM WITH 2D SEISMIC DATA

- 2D seismic data only cover thin slice of the sub-surface.
- There are problems associated with off-line reflections. Off-line reflections are sometimes referred to as sideswipe and are indistinguishable from the reflections directly below.

Sideswipe is a type of event in 2D seismic data in which a feature out of the plane of a seismic section is apparent, such as an anticline, fault, or other geologic structure. A properly migrated 3D survey will not contain sideswipes.

The only way to solve both coverage and sideswipe problems is to cover the entire survey area with a series of very closely spaced seismic lines, producing a 3D volume of seismic data.

3D SEISMIC DATA

In 3D survey, shot is within the grid of multiple lines of receivers and the next line is spaced tens of metres away. In other words, a 3D seismic volume is created by shooting a closely spaced grid of 2D lines and interpolating between the lines to create a 'three-dimensional volume' of data that is also referred to as a cube (Figure 3.23). Both land and marine 3D data are acquired with multiple source and receiver arrays to facilitate the acquiring of large volumes of data (Figure 3.22).

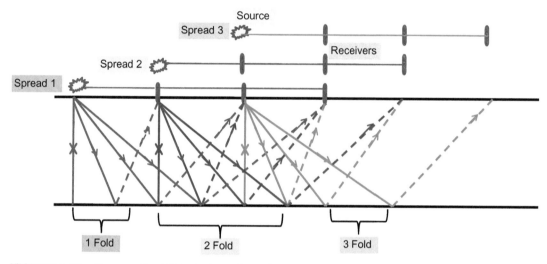

FIGURE 3.22   Conceptualized 3D seismic configuration.

BENEFITS OF 3D SURVEY

- 3D seismic method often improves data density
- Resolves many of the problems found in 2D sections, such as out-of-plane reflections or sideswipe
- 3D seismic data provide detail information about fault distribution and the sub-surface structure unlike 2D
- Because 3D seismic method provides a cube of data that represent a volume of the earth, it allows us to examine data in many different ways. The results are an improved understanding of structures and nature of the earth beneath us and enhance probability of finding recoverable reserves of hydrocarbon.

FIGURE 3.23   3D seismic cube. *Source: Clifford, A., Goodman. E., 2010. Southeast Louisiana shallow gas – 1: Louisiana lagniappe: shallow gas play concept, evaluation techniques, analogs. Oil Gas J.*

4D (3D-TIME LAPSE) SEISMIC SURVEYS

4D seismic survey is a three-dimensional (3D) seismic data acquired at different times over the same area to assess changes in a producing hydrocarbon reservoir with time. Changes may be observed in fluid movement and saturation, pressure, and temperature.

The oil and gas industry uses 3D-time-lapse seismic survey to monitor the way fluids flow through a reservoir during production, by carrying out a baseline (pre-production) seismic survey (Figure 3.24) and then repeat surveys over the production lifetime of the reservoir (Figure 3.25). When 3D surveys are repeated in this way, they are often referred to as 4D seismic.

Typically, 4D seismic data are processed by subtracting the data from the baseline 3D survey from the data from the monitor 3D survey. The amount of change in the reservoir is defined by the difference between the two. If no change has occurred over the time period, the result will be zero.

FIGURE 3.24 Conceptualized baseline 3D seismic survey.

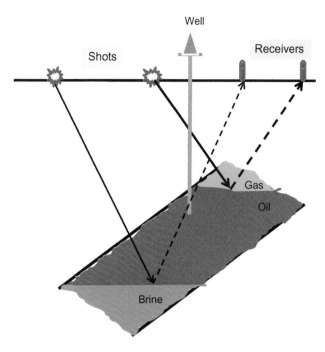

FIGURE 3.25 Conceptualized monitor 3D seismic survey (after x years oil production).

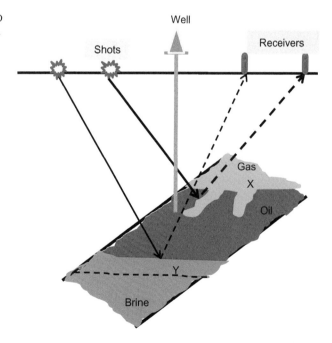

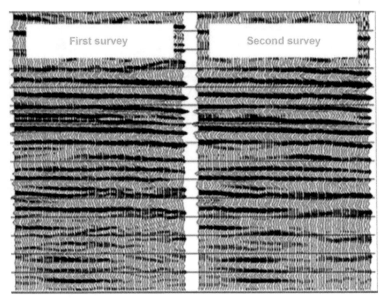

FIGURE 3.26   4D seismic survey data. *Source: FairfieldNodal. www.fairfieldnodal.com.*

Taking a closer look at both surveys in Figure 3.26, you will notice a remarkable difference between both surveys. This difference is called the 4D signature.

BENEFITS OF 4D SURVEY

- 4D survey will allow the reservoir engineers and geoscientists using state-of-the-art computer systems to optimize static and dynamic models of the complex reservoir. The end result of such a survey will be enhanced recovery of hydrocarbon from the field, better sitting of production and injection wells, and reduced cost of drilling and prolonged life of the field.
- 4D survey provides information for efficient management of oil and gas field. For example, to monitor the expansion of a gas cap, with the intention of managing production so as to prevent it from reaching producing wells and thus decreasing the oil production rate.
- It also provides a baseline for the development of other oil and gas fields that have been identified in the vicinity.
- 4D survey can help to locate untapped pocket of oil and gas in a reservoir.

SIMILARITY BETWEEN THE BASELINE AND THE MONITOR 3D SURVEYS

At the top of the reservoir, the two-way time between the baseline 3D survey (pre-production) and the monitor 3D survey is the same.

DIFFERENCE BETWEEN THE BASELINE AND THE MONITOR 3D SURVEY

In the reservoir (Figure 3.25), the oil has decreased by Y as oil is being produced from the well, and the gas expanded by X. This is because at the producing well, the pressure drops. As the pressure decreases further, gas will come out of solution. This will result in decrease in seismic velocity and density and therefore decrease in acoustic impedance. Therefore, the two-way time thickness of the reservoir will change as oil is produced.

Seismic events below the reservoir will therefore change in two-way time between the baseline and the monitor 3D seismic surveys. Such time-shift in and below the reservoir gives rise to the time-lapse effect that is noticeable between the baseline and the monitor 3D surveys.

Monitor 3D seismic survey is different from the baseline 3D survey in that it adds the element of time so that changes in the fluid composition in the reservoir can be monitored over the production life of the field (Figure 3.27).

FIGURE 3.27 Changes in a reservoir seen through 4D seismic. *Source: Schlumberger.*

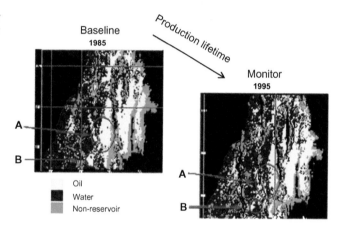

Note that differences in seismic amplitudes or travel times between the monitor and baseline surveys can reveal the movement of fluids or the extent of pressure changes that affect reservoir properties.

Also, the ambient noise for the baseline 3D survey will be different from the monitor 3D survey. This is because as production takes place there will be many sources of noise generated by production facilities that were absent when the baseline survey was shot.

Furthermore, some source areas that were shot in the baseline (pre-production) survey cannot be shot in the monitor 3D survey during production because of production facilities such as the FPSO (floating production storage and offloading) and SPM (single point mooring) in the area. The FPSO and the SPM will prevent the monitor survey vessel from acquiring data in that area and this will create a hole in the acquired data (white area in Figure 3.28) and makes imaging the sub-surface reservoir challenging. Therefore, it will not be possible to acquire a monitor 3D survey that is the same as the baseline 3D survey.

OCEAN BOTTOM NODE (OBN)

OBN technology is introduced to enable surveying in areas of obstructions (such as production platforms) or shallow water inaccessible to ships towing seismic streamers (floating cables).

As stated earlier, 3D monitor streamer surveys tend to leave seismic holes in under-platforms areas or in areas inaccessible to towed streamers. An image of this is shown in Figure 3.28.

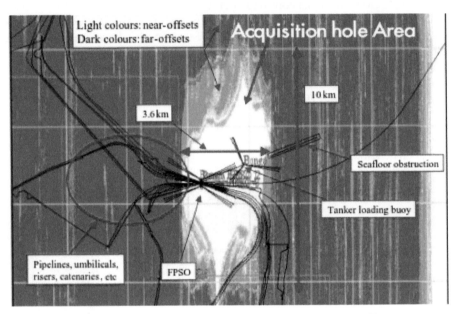

FIGURE 3.28   An offshore field with seismic hole in near-offset data around the area where the production facilities are located from a 3D streamer survey. *Source: NAPE extended abstract 2011.*

In other words, 3D streamer surveys lack near-offset data and poorly imaged the producing reservoirs. To mitigate this effect and improve sub-surface reservoir imaging and interpretation, OBN (Ocean Bottom Node) technology is employed as seismic infill to an earlier 3D monitor streamer survey around the production facilities.

In the OBN survey, every receiver location is a 4C sensing system: a three-component geophone and a hydrophone.

OBN is acquired using two vessels: a source vessel and a nodal vessel.

OBJECTIVE OF OBN SURVEY

- To improve the image of the producing reservoirs underneath the production facilities.
- To provide seamless data acquisition and processing that can enhance and combine with existing 3D streamer, OBC, or permanent reservoir monitoring programmes to provide a more complete view of the sub-surface for enhanced exploration, field development, and reservoir optimization.

OBN SEISMIC SOURCE

Airgun is used as the seismic source to generate seismic energy. The airgun used has an operating pressure of about 2000 psi, nominal operating airgun depth is 10 m, and gun volume 4370 in.^{3}, at a pop interval of 18.75-m dual source. An airgun array (Figure 3.29) are use to improve effectiveness as well as to makes it a better source.

While the receiver system is stationary on the seabed, a source vessel towing a marine source array shoots on a pre-determined dense grid on the sea surface.

FIGURE 3.29   An array of airgun whose operating pressure is 2000 psi. *Source: www.offshoreenergyresearch.ca.*

RECEIVER

In OBN survey, every receiver location uses a 4C component sensing system. That is, three orthogonally oriented geophones and one hydrophone with an ocean-bottom sensor (deployed in node-type systems as well as cables). The receivers (geophones and hydrophones) are place in contact with the seabed. The geophones allow for the recording of S-waves, while the hydrophone measures P-waves and can be fixed in position to allow for repeatable seismic records, which is important for reservoir monitoring.

The nodes are place on the seabed with the aid of a remotely operated vehicle (ROV). The ROV is fitted with a custom built tool known as the Beck Tool, which is used to deploy the

FIGURE 3.30   The node been placed on the seabed using ROV. *Source: SUB-SEA world news – The Netherland: Fugro Awarded Ocean Bottom Node Survey Program. Posted on 5 September 2012. www.subseaworldnews.com.*

Remotely operated vehicle (ROV)   Node on the seabed

sensor (node), containing the hydrophones and geophones, and plants it into the seabed. This is shown in Figure 3.30.

Note that the nodes use a battery that can last for 65–70 days.

When the shooting vessel has acquired sufficient source lines, the nodes are retrieved, data downloaded, and the nodes are redeployed into the seabed in a nearby location and the acquisition work continues.

USES OF OBN SURVEY

The uses of OBN data can be divided into three broad categories:

- Firstly, lithology and fluid prediction by the combined analysis of pressure and shear waves.
- Secondly, time-lapse (4D) seismic monitoring.
- Thirdly, imaging in geologically complex areas.

COMPARISON BETWEEN OBN AND OBC (3D STREAMER SURVEY)

OBN survey can provide unique data quality and therefore, more information can be obtained for improving characterization of subsurface compares to 3D streamer data.

OBN acquisition offers the prospect of full illumination and high multiplicity of signals from the same sub-surface points (high data fold).

Although conventional 3D streamer seismic data serve exploration purposes well in many cases, the data quality may not be sufficient to support an adequate model for reservoir development, in particular below complex overburden. This experience led to the development of OBN technology.

1996 OBC data   2004 OBN data

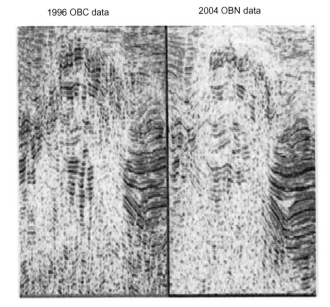

FIGURE 3.31 Imaging comparison of 1996 OBC (3D streamer survey) and 2004 OBN data at Cantarell. *Courtesy: Pemex/SeaBed Geophysical. Source: Amundsen, L., Landro, M., 2008. Seismic imaging technology part 111. GeoExPRO Magazine 5(4). www.geoexpro.com.*

OBN uses 4C receivers that are placed directly on the seafloor by a remote operating vehicle (ROV), properly GIS referenced and therefore allow the most accurate positioning and orientation.

Advantages of OBN systems are their high repeatability in 4D applications for reservoir monitoring, their ability to operate in complex topography or infrastructure areas, and their excellent coupling with the seafloor compared to 3D streamer surveys.

Note the drawback in OBN is that nodes cover a sparsely sampled acquisition zone.

From Figure 3.31, comparing the OBC 1996 data, with the Seabed's OBN 2004 data, the OBN data demonstrated higher resolution, excellent reflector continuity, and improved structural definition.

RECORDING SEISMIC DATA ON TAPE

This flow chart show the key steps involved with recording and handling seismic data.

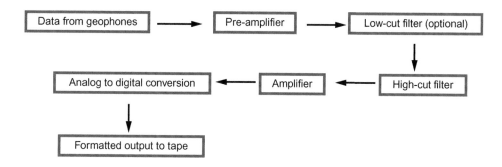

The geophones or hydrophones measure the seismic wave field, and their output is electrical signals. The voltages, that is, the data from the geophones are initially amplified at the preamplifier and may then pass through analogue low-cut filter (a low-cut-off filter applied to a seismic signal attenuates frequencies less than the cut-off frequency).

The data are filtered by the analogue high-cut filter (a high-cut-off filter applied to a seismic signal attenuates frequencies above the cut-off to make the output more low frequency) so that frequencies above Nyquist are removed and then amplified again.

Note that the maximum frequency beyond which aliasing will occur is called the Nyquist Frequency.

Finally, the data are converted from analogue to digital information, before being recorded on tape in a known readable format.

RECORDING MULTI-CHANNEL SEISMIC DATA

Multi-channel seismic data recording is the process of recording the reflections from a single seismic shot with multiple channels (that is, groups of geophone in a string) at the surface. Such redundancy of recorded data enhances the quality of seismic data when the data are stacked

ISSUES WITH RECORDING MULTI-CHANNEL SEISMIC DATA

- The first issue is that as soon as the seismic shot (Figure 3.32) is fired all of the receiver groups will be receiving data simultaneously. If these data are recorded on tape as soon as it is available, the data on tape will be in the wrong order. This order is not appropriate for later processing of the data. The geophysicists will have all data at one time sample, followed by all data at the next sample – the data are ordered by time, not as seismic traces. This type of recording format is known as multiplexed recording.

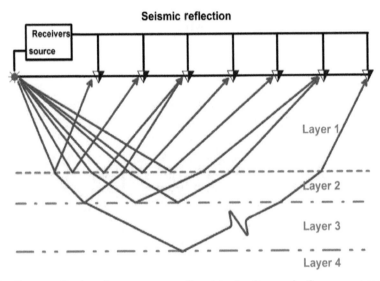

FIGURE 3.32　Conceptualized receiver groups recording data simultaneously. *Source: www.enviroscan.com.*

- Second issue is that the collection of data is done in time-sequential order. The first sample of channel 1 is collected followed by the first sample of channel 2, then the first sample of channel 3, etc., until the first sample of all the channels are collected. Then the second sample of channel 1 is collected and then followed by the second sample of channel 2, etc. There is a slight time difference between the samples of the channels, as data from later channels are recorded later than data from earlier channels. This multiplexer skew, dependent on the channel number, is usually corrected in the earliest stages of data processing – when the data are being de-multiplexed from the field tape into an internal tape format in trace order.

DIFFERENCE BETWEEN MULTIPLEXED AND DE-MULTIPLEXED

Multiplexed Data Format

When data are stored in multiplexed format, the groups of numbers are referred to as a scan. This is one time sample for all channels, or geophone groups, in channel number order.

De-Multiplexed Data Format

When data are stored in de-multiplexed format, the order is changed so that all samples for one receiver group or channel are kept in time order; this is referred to as a trace.

In other words, de-multiplexed data format is a trace-sequential format. The first sample of channel 1 is collected followed by the second sample of channel 1 and then the third sample of channel 1, etc., until all of channel 1 is collected, followed by the first sample of channel 2, etc.

Note that the difference between multiplexed data format and de-multiplexed data format is simply in the order in which one field record or shot is stored.

Table 3.1 shows the numbers recorded in a seismic record. Each row is recorded at one time; each column is recorded from one channel. The multiplexed data order would be D001, E001, and F001. The de-multiplexed data order would be D001, D002, and D003.

TABLE 3.1   Conceptualized how Multiplexed and De-Multiplexed Data are Ordered

Time	Channel		
	1	2	3
0.001	D001	E001	F001
0.002	D002	E002	F002
0.004	D003	E003	F003

STANDARD WAY SEISMIC DATA IS STORED

Every processing contractor and Oil Company has their own format for storing seismic data during in-house processing of the data. But the society of exploration Geophysicists has a standard for storage of acquired and processed seismic data. Seismic data are stored on tape either in SEG-D or SEG-Y format.

SEG-D Format

SEG-D is used to store acquired seismic field data. It can contain multiplexed or de-multiplexed data.

Multiplexed data are stored in a series of scans per block. The set of data values of each scan is preceded by a Start–Of–Scan code and time value, which identifies the scan.

De-multiplexed data are stored in a sequence of a header record followed by a number of de-multiplexed data records and then a single end file mark. This sequence is repeated for each shot on the line.

SEG-Y Format

SEG-Y is a trace-sequential (or de-multiplexed) format used for storage of processed seismic data or partially processed data from field processing crew. SEG-Y data format enhances data transfer from one contractor to another. That is, SEG-Y is used for data exchange.

All tape formats share a common general layout. Each tape or file on a multi-file tape consists of some or all of the following:

- A general header that identifies the tape
- A header for each 'batch' of data, for example, one field record on the tape
- A header for the individual traces or scans on the tape
- The seismic samples

Note that a file may consist of many individual traces or records, and the tape may consist of many different files. Each file on the tape is ended by a particular 'End of File' code. Then the recording sequence is repeated along the length of the tape until the tape is full.

Understanding Noise
in Seismic Record

The aim of seismic data acquisition is to obtain data that contain only primary, reflected waves. But this is not possible because the generation of unwanted waves is inevitable during seismic data acquisition. Any rays in a seismic record other than primary reflections are called noise.

TYPES OF NOISE ON SEISMIC RECORD

There are two kinds of noises found in seismic data: random noise and source-generated noise.

RANDOM NOISE

Random noise is noise generated by activities in the environment where seismic acquisition work is being carried out. In a land acquisition, random noise can be created by the acquisition truck, vehicles, and people working in the survey area, wind, electrical power lines, and animal movement. This noise appears in a seismic record as spikes.

In a marine acquisition, random noise can be created by ship props, drilling, other seismic boats, and wind/tidal waves.

63

Radom noises in seismic data are recognized principally by the absence of coherency or continuity from one seismic trace to the next (Figure 4.1). Note that coherent noise is consistent from trace to trace.

Random noise can be reduced or removed from data by stacking the traces, filtering during processing or using arrays of geophones during acquisition.

SOURCE-GENERATED NOISE

Source-generated noise is created by the seismic source. Source-generated noises are coherent noise trains and they exist in an organized form from trace to trace and yet they contain no geologic information. Source-generated noises are in the form of surface wave which exhibit strong coherency and essentially obscure the entire desired primary reflection. Source-generated noises noticed in seismic records are ground roll, direct rays, ghost reflections, and multiple reflections.

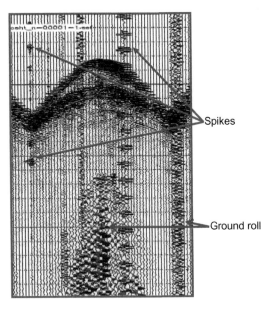

Spikes

Ground roll

FIGURE 4.1   Noise in the seismic record. *Source: Shell E&P.*

Surface Waves

Other than body waves, the other type of seismic waves is surface waves. Surface waves move along the surface of the earth and they are of lower frequency than body waves (P-waves and S-waves). Surface waves are Rayleigh wave and love wave.

Ground roll (Rayleigh wave) is a type of coherent noise observed in seismic field record. Ground roll occurs as a set of dispersed wave trains with low velocity, low frequency, and high amplitude. Ground roll can obscure the primary reflected events and degrade overall data quality, but can be alleviated through careful selection of source and geophone arrays, filters, and stacking parameters.

Note also that longer wavelength components of the ground roll penetrate more deeply in the surface layer than the shorter wavelengths and thus see higher velocities. Longer wavelength means lower frequency.

Since

$$f = \frac{v}{\lambda}$$

where f is the frequency, v is the seismic velocity, and λ is the wavelength.

Ghost Reflections

Ghost reflections are produced when the seismic shot first reverberates upwards from the shallow subsurface and reflects downwards and travel a similar path to the "normal shot ray" from the shot to the reflector. The ray path returning from the reflector does exactly the same thing at the receiver and joins the primary reflected ray path from the receiver in a similar way (Figure 4.2).

FIGURE 4.2   The figure shows how ghost reflection is formed.

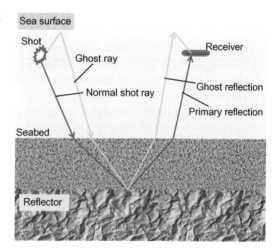

When ghost reflections are formed, the original signal detected by the receiver will undergo interference from the ghost signal. Depending on the frequency of the signal and the relative distances travelled by the two signals, the effect of the interference will vary.

Note that interference is the effect produced when two waves of the same frequency, amplitude, and wavelength travelling in the same direction in a medium are superposed – as they simultaneously pass through a given point.

Multiple Reflections

Multiple reflections are produced when energy from the seismic shot travels down to the seabed, then up to the surface, reflecting multiple times before travelling a similar path from the shot to the reflector. The ray returning from the reflector does exactly the same thing at the receiver (Figure 4.3).

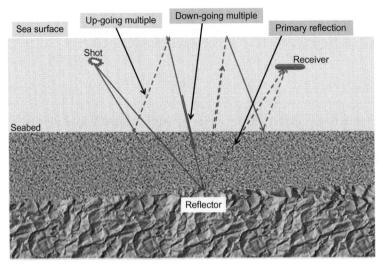

FIGURE 4.3   Up-going and down-going multiple reflections.

There are paths in which the acoustic energy is reflected by a deeper reflector, then reflected again by a shallower one, subsequently reflected again by a deeper one. The wave traverses the same layer of the earth multiple times, and the energy may reflect more than once on any reflector.

Depending on their time delay from the primary events with which they are associated, multiples are characterized as short path, implying that they interfere with the primary reflection, or long-path, where they appear as separate events from the primary reflections.

The water layer is one of the primary causes of multiple reflections and any layer with sufficiently strong acoustic impedance contrast can become a source of multiple reflections on both land and marine data.

In marine, the air–water interface and the water–bottom interface allow a sequence of strong reverberations to occur. These appear with regularity at successively later times and can obscure the simultaneously arriving primary reflection events.

The amplitude of the multiples will normally be sign-reversed due to the sea-surface reflector and will appear as an image of the primary reflector, usually some constant time below it.

Further Reading

Chouch, A., Mari, J.L. 3-D land seismic surveys – definition of Geophysical parameters.

Trilobit 4C ocean bottom nodes (fig. 23). Source: A favourable outlook for high-end seismic, vol. 232(9), September 2011. WorldOil Online magazine. www.worldoil.com.

Applied seismology: a comprehensive guide to seismic theory and application by Mamdouh R. Gadallah and Ray L. fisher.

Barry, K.M., Cavers, D.A., Kneale, C.W., 1975. Recommended standards for digital tape formats. Geophysics. 40 (2), 344–352. http://dx.doi.org/10.1190/1.1440530 Bibcode: 1975Geop. . .40..344B.

Basic Petroleum Geology, Halliburton, 2001.

Basic Exploration Seismology, Robinson & Coruh.

Basin Principle in Tectonic (Fig. 1.13). Universidada Fernando Pessoa Porto, Portugal. Homepage.ufp.pt.

2010. Imaging the invisible – BP's path to OBS nodes. SEG Expanded Abstracts (SEG). 29, 3734.

Beaudoin, G., Reasnor, M., 2010. Atlantis time-lapse ocean bottom node survey: a project team's journey from acquisition through processing. SEG Expanded Abstracts (SEG). (29), 4155.

Fundamentals of Petroleum Geology by Prof. Etu-Efeotor, J.O.

Gerding, Mildred (Ed.). Fundamentals of Petroleum, third ed. Petroleum Extensive Service, Division of Continuing Education, The University of Texas at Austin, Austin, Texas, USA, (1986). ISBN 0-88698-122-0.

Gausland, I., 2000. Impact of seismic surveys on marine life. The Leading Edge (SEG). 904 Retrieved 8 March 2012.

Geology Time Scale: Explore Montana Geology. www.mgmb.wtech.edu.

Johnson, M.G., Gaskins, G.M., Greenlee, S.M., 1994. 3-D seismic benefits from exploration through development: an exxon perspective. Leading Edge. 13, 730–734.

Howe, D., et al., 2008. Independent simultaneous sweeping – a method to increase the productivity of land seismic crews. SEG Expanded Abstracts (SEG). 27, 2826.

http://au.answers.yahoo.com/question/index.

http://wiki.answer.com/seismic.

http://www.cfhd.gov/agm/images/fig.118.jpg.

http://www.cggveritas.com/default.aspx?cid=4662&lang=1.

http://www.geol.isu.edu/faculty/juan/reflectseismo197/rcbradley/wwww/rcbradley1.html.

http://www.sintef.no/Projectweb/co2fieldlab/test-methods/what-is-seismics/.

http://en.m.wikipedia.org/wiki/structural_trap.

http://en.m.wikipedia.org/wiki/Thrust_fault.

http://www.geosci.usyd.edu.au.

Introduction to Petroleum Seismology by Luc T. Ikelle and Lasse Amundsen.

Bacon, M., Simm, R., Redshaw, T., 2003. 3-D Seismic Interpretation. Cambridge University Press. ISBN 0-521-79203-7.

McCauley, R.D., et al., 2000. Marine seismic surveys: a study of environmental implications. APPEA. 692–708 Retrieved 8 March 2012.

Nestvold, E.O., Jack, I., 1995. Looking ahead in marine and land geophysics. The Leading Edge. 14, 1061–1067.

Anderson, Richard G., McMechan, George A., 1988. Noise – adaptive filtering of seismic shot records. Geophysics. 53 (5), 638–649.

Northwood, E.J., Weisinger, R.C., Bradley, J.J., 1967. Recommended standards for digital tape formats. Geophysics. 32 (6), 1073–1084.

Norris, M.W., Faichney, A.K. (Eds.), 2002. SEG Y rev1 Data Exchange Format. Society of Exploration Geophysicists, Tulsa, OK.

Kearey, Philip, Brooks, Michael, Hill, Ian, 2002. An Introduction to Geophysical Exploration. Blackwell Science Ltd. ISBN 10: 0632049294.

Oil & Gas Science & technology – Rev, 2006. IFP 61(5).

RIGZONE – How Do 4-D and 4-C Seismic work? www.rigzone.com.

Schlumberger Oilfield Glossary. Ground Roll. http://www.glossary.oilfield.slb.com/Display.cfm?Term=ground%20roll.

Schlumberger Oilfield Glossary. OBC.

Schlumberger Oilfield Glossary. Four-Component Seismic Data.

Schlumberger Oilfield Glossary. Multiple Reflection.

Seismic field record (fig. 1.53). Source: GEOSPHERE INC – seismic methods: Data Processing. www.geosphereinc.com.

Seismic Imaging Technology Part iii by Lasse Amundsen and Martin Landro (fig. 1.72). Geo ExPRo magazine, issue 4, volume 5, 2008. www.geoexpro.com/article/seismic_imaging_Technology-partiii.

Sheriff, R.E., 1991. Encyclopedia Dictionary of Exploration. Society of Exploration Geophysics, Tulsa, OK.

Sheriff, R.E., Geldart, L.P. (Eds.), 1995. Exploration Seismology, second ed. University of Houston, Cambridge University Press. ISBN 9780521468268.

Simm, R., White, R., 2002. Phase, polarity and interpreter's wavelet. First Break. 20, 277–281.

Structural Geology of Sedimentary Basins: Sedimentary basin extension zones by Philip Hutson, James Middleton, Daniel Miller & Adams Wallenstein.

SUBSEA World news (fig. 1.71) – The Netherlands: Fugro Awarded Ocean Bottom Node Survey Program. Posted on 5 September 2012.

Telford, W.M., Geldart, L.P., Sheriff, R.E., Keys, D.A., 1976. Applied Geophysics. Cambridge University Press, Cambridge, UK.

Three-dimensional seismic data: Schlumberger Oilfield Glossary.
www.glossary.oilfield.slb.com.
UKOL. University of Derby. www.sub-surfrocks.co.uk/seismicexp.html.
Unspooling miles of seismic streamer near the Shumagin Island by Donna Shillington. The Earth Institute Columbia
    University. Blogs.ei.columbia.edu.
www.google.com.ng/imres?q=seismic+airgun&num.
www.google.com.ng/imgres?q=seismic+vessel.
www.google.com.ng/imgres?q=ocean+bottom+node&num.
www.google.com.ng/imgres?q=seismic+streamer&num.
\www.google.com.ng/imgres?q=3D+seismic+cube&start.
www.google.com.ng/imgres?=seismic+recording+tape&num.
www.offshore-technology.com/contractors/rov/oceanor/.html.
www.geo.mut.edu/UPSeis.
www.geo.mtu.edu/UPseis/waves.html.
www.geosci.usyd.edu.au/.../Listric_Faults.
http://www.glossary.oilfield.slb.com/Display.cfm?Term=multiple%20reflection.
Stewart, J., Shatilo, A., 2004. A comparison of streamer and OBC seismic data at Beryl Alpha field, UK North Sea. SEG
    Expanded Abstracts (SEG). 23, 841.
3D seismic cube. Source: SOUTHEAST LOUISIANA SHALLOW GAS – 1: Louisiana lagniappe: shallow gas play
    concept, evaluation techniques, analogs by Andy Clifford and Elizabeth Goodman. OIL&GAS JOURNAL,
    12/06/2010. www.ogj.com.
Omudu, L.M., Suleiman, I., Segun, O., Olambiwonnu, R., Ejiofor, A., Quadt, E., Mbah, R.O., Olotu, S., Osayande, N.
    Bonga Ocean Bottom Nodes (OBN) Seismic Survey: Acquisition Challenges (fig. 3.28). Shell Petroleum Develop-
    ment Company of Nigeria Port Harcourt Nigeria. NAPE Extended Abstract 2011.

DETAILED SEISMIC DATA PROCESSING TECHNIQUES

Understanding the Detail Seismic Processing Techniques Used to Convert the Acquired Seismic Data into the Geologic Section of the Earth

WHAT IS SEISMIC DATA PROCESSING

Seismic processing is the alteration of the acquired data to suppress noise, enhance the recorded seismic trace and migrate the seismic trace to its correct location in space and time. Processing steps include static corrections, deconvolution, normal moveout, velocities analysis, dip move-out, CMP/CDP stacking and migration, which can be performed before or after stacking.

The objective of seismic data processing is to remove all noise and distortions introduced by the seismic acquisition method and produce a seismic section as close as possible to the subsurface image of the earth that can be interpreted as shown in Figure 5.1.

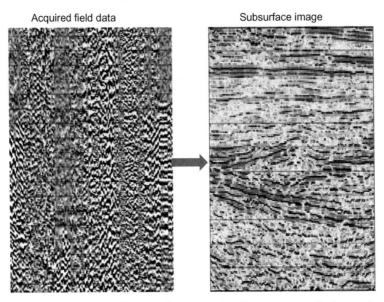

FIGURE 5.1   Described the objective of seismic data processing. *Source: www.geophysics.geoscienceworld.org.*

The data/information required to extract the subsurface image (geologic section of the earth) from the acquired seismic data are as follows:

- Field seismic data in SEG-D format – it does not contain coordinate information of each shot and geophone group (channel).
- The (X, Y, Z) coordinates for each shot and geophone group and this is obtained using global positioning system (GPS). The GPS allows the surveyor to produce an accurate map showing every shot and geophone position in the surveyed area.
  Note that wrong shot and geophone group coordinates will lead to mis-positioning and distortion of the final seismic image (migrated image).
- Accurate elevation information of every shot and geophone groups (for land data). The geophone group is also known as geophone station.
- Information on which geophone groups are "live" for each shot.

TRANSCRIPTION

The acquired field data are stored in SEG-D or SEG-Y (if it has been processed in the field) format.

Transcription is the process whereby all of the field data are converted into the company's internal storage format in the initial stage of seismic data processing.

If the data have been recorded in a multiplexed format (ordered by channel, not trace), then the computer must be able to read the entire field record in one go and then reorder it (de-multiplexed).

A field record consists of all trace samples recorded in a single shot. After de-multiplexing the geophysicists check to ensure that every trace of every shot has been read correctly.

OBSERVER LOG

The observer log contains information about the acquired seismic field data. The geophysicists usually compare the list of shots read from the tape against those listed in the observer logs. A field record number may be used which is designated to match the shot-point number and could get out of step.

A scan error may occur. A scan error is when one multiplex scan of the channels gets out of "sync". This means that the shot will have to be omitted from any processing, but the geophysicists usually look at the data as soon as possible to confirm this. The geophysicist checks this initial stage to ensure that every trace of every shot has been read correctly.

SPHERICAL DIVERGENCE

As the acoustic wave emitted by the seismic source travel through the subsurface, its energy expands in all direction like a sphere (Figure 5.2). All the energy initially contained in the seismic source is spread out over a wider area as time increases. This causes loss of energy in the seismic signal and resulted to decrease in the amplitude of the source wavelet and is referred to as spherical divergence.

By definition, spherical divergence is the apparent loss of energy from a source wavelet as it propagates through the subsurface. Spherical divergence decreases energy with the square of the distance.

Such energy loss must be accounted for when restoring seismic amplitudes to perform fluid and lithology interpretations, such as amplitude versus offset (AVO) analysis.

FIGURE 5.2 Conceptualized spherical divergence.

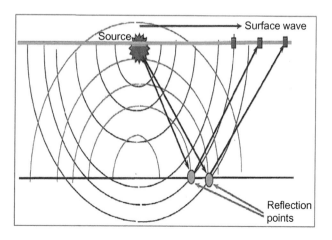

In a constant velocity medium, amplitude losses caused by geometrical spreading (spherical divergence) can be corrected for by multiplying by time t. But rock velocities are not constant and the rate at which seismic energy expands depends on the velocity of the rock through which it is passing through. So actual wave fronts are not spherical, and their area increases at a faster rate than in spherical divergence.

Energy losses on a shot record such as spherical divergence and inelastic attenuation (adsorption and scattering) influence the wave amplitude and reflection coefficient. But the geophysicists want to preserve true seismic amplitude because they contain information about the subsurface geology. In other to balance the seismic data and remove these amplitude effects, a general amplitude modification process is applied to balance the data at all time and it is called gain function.

Note: A gain is the change in the amplitude of an electrical signal from the original input to the amplified output. It is this concept in electronics that we are applying in seismic to improve the visibility of the recorded seismic trace.

True amplitude processing are the steps used in seismic processing to compensate for attenuation, spherical divergence and other effects by adjusting the amplitude of the data. The goal is to get the data to a state where the reflection amplitudes relate directly to the change in rock properties giving rise to them.

Since the rate at which seismic wave propagates through a rock depends on the velocity of the rock through which it has travelled, the geophysicists usually have no idea what the velocity is. Therefore, the seismic data themself are used to establish a function that is used to compensate for the energy loss.

AUTOMATIC GAIN CONTROL (AGC)

AGC is a system that controls the increase in the amplitude of an electrical signal from the original input to the amplified output, automatically.

AGC is used in data processing to improve the visibility of seismic data in which attenuation or spherical divergence has caused amplitude decay (Figure 5.3).

Data without AGC   Data with AGC applied

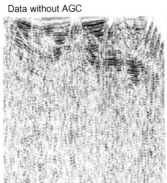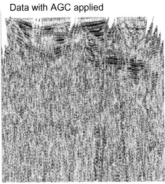

FIGURE 5.3 Seismic data without AGC (left) and the same data after AGC is applied. *Source: crack.seismo.unr.edu.*

You can notice that the energy is not visible in some of the record on the left of Figure 5.3. AGC applied approximately removes the loss of energy (spherical divergence) and equalized the amplitudes with improved resolution (on the right). AGC makes all events visible and appear balance in the data, but relative amplitude (AVO) information is lost.

In seismic processing, RMS AGC is used for amplitude equalization. A key parameter in RMS AGC is the gate length. The RMS is simply the square root of the average amplitude squared in the window. It gives the geophysicists a measure of the overall absolute amplitude in the window, both as positive and as negative values. Note that the "window" is the length of the seismic data that AGC is applied on.

AGC can be fast or low. A very short gate is similar to fast AGC and resulted in elimination of almost all amplitude variation.

Note also that amplitude variation caused by subsurface geology must be preserved because they tell the geophysicists much about certain qualities of the subsurface.

In seismic data processing, the geophysicists avoid any steps that will cause amplitude artefacts. Though they would prefer to have data where amplitude is everywhere proportional to reflectivity, this is not possible. The geophysicists can only have seismic data where lateral variation of amplitude on a particular group of reflectors is proportional to reflectivity changes.

The geophysicists assume that the average absolute reflectivity over a long time window varies little, so a long-gate AGC can be applied to the data. The benefit of long gate is that it avoids destroying the lateral variation in amplitude. The geophysicists also ensure that the gate include many reflectors so that the target event makes very little contribution to the average amplitude in the window.

Note that effects like bright spots will usually survive the application of AGC, but subtle features such as AVO must be carefully preserved.

For structural imaging/interpretation, automatic gain control (AGC) is often applied. This preserves the root-mean-square (RMS) amplitude in a user-defined window.

REFRACTION STATICS

Seismic data recorded on land do not follow regular pattern. Due to the elevation and topography of the surface, it may be impossible to arrange the shots and geophones in a straight line (Figure 5.4).

FIGURE 5.4 Conceptualized variation in thickness in the weathered layer.

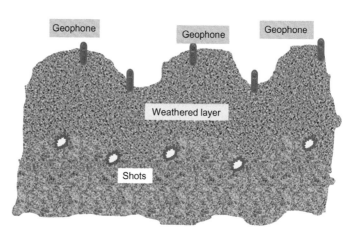

If shots and geophones are not in the same elevation, then the line-up of reflections will be distorted. The arrival of the reflected wave will be delayed when the wave has to travel through more rocks. As a result, the line-up of a flat horizon on a shot record will appear to be distorted due to the different travel times of the rays, which originated from the shot and were recorded with receivers at different elevations (T_0 in Figure 5.5).

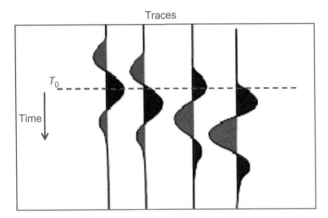

FIGURE 5.5   The effect of variable velocity in the weathered layer.

Similarly, the line-up of such a reflector on a receiver record will be distorted due to the different travel times of the acoustic waves from the various shots which are recorded with one receiver. So the geophysicists need accurate elevation information of every shot and geophone groups for correct processing of the acquired seismic data.

Also, the layer just below the earth surface is the weathered layer. It has variable seismic velocity and thickness. The weathered layer is the low-velocity layer near surface in which rocks are physically, chemically or biologically broken down. That is, the low-velocity layer consists of unconsolidated material.

Velocity variation in the near surface can occur horizontally as well as vertically. Along the length of one seismic line, the near-surface thickness can vary as much as 100 m, and its velocities as much as 900 m/s.

Gradual velocity variation in the near surface introduces apparent structure and dipper reflections on the seismic section. Rapid changes in velocity in the near surface introduce time differences within a CMP gather. Thus, the variable seismic velocity and thickness in the near surface causes trace-to-trace time differences, as in Figure 5.5, which gives false picture of the subsurface. If it were not for the variation in the weathered (near-surface) layer, all events would have been recorded at time T_0 in Figure 5.5. These variations which cause reflection times variations must be corrected for.

STATICS CORRECTION

Statics correction is a bulk time shift applied to the recorded seismic trace during data processing.

Static corrections are applied to remove the effect of variation in the weathered layer and any changes in the near surface that causes trace-to-trace time difference. These corrections are based on up-hole data, refraction first breaks or event smoothing.

In other to calculate the time corrections (statics) for every shot and geophone stations (geophone groups), it is necessary to look at ray paths, which occur, in the weathered layer.

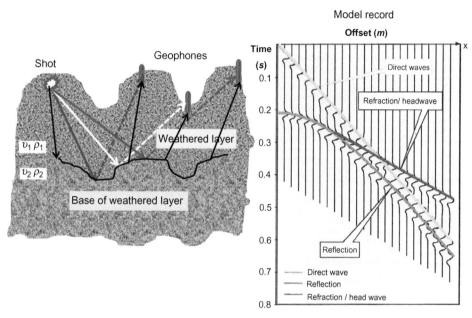

FIGURE 5.6   Ray paths (left) and T–X plot (right). *Source: NAPE summer sch. 08 slide.*

Three types of events are recorded on the early part of the seismic field records. They are as follows:

- Direct waves.
- The reflected waves that travel at velocity V_1 and reflecting from the boundary at the base of the weathered layer.
- Refracted waves, reaching critical angles and travelling along the base of the weathered layer.

Note that at the lowest offset (right of Figure 5.6), no refracted wave is formed but the first arrival is formed by the direct wave, followed by the reflected wave. At the other offset, the order of arrival changes as each wave moves at its own velocity.

Figure 5.7 shows the first part of a seismic record (shot gather) that is used to establish the low-velocity information, which is needed to calculate for static correction.

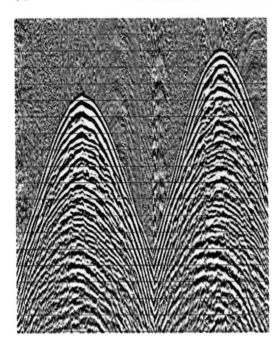

FIGURE 5.7   Shot gather.

Static corrections are calculated by picking the first arrivals. It is usually auto-picked by a computer.

By definition, first arrival or first break is the first indication of seismic energy on a trace. On land, first breaks represent the base of weathered layer and are useful in making static corrections.

Figure 5.8 shows the picked event (first arrivals). The picks event is used to calculate the depth of the weathered layer by analyzing the velocities of the refracted arrivals (red line in the T–X plot in Figure 5.6). By picking the horizontal velocities apparent within the first breaks, the geophysicists are able to establish estimates of the shallow velocities. Once the velocities have been identified, the intercept times are determined by extrapolating the lines back to zero offset. It is this information that is used to establish the depths of the weathered layers.

Note that the picking is done for every trace on every shot of every line. It is the time difference between successive traces that is important and random errors in absolute timing are canceled out in the complex calculations done on the result.

The depth of the weathered layer is calculated using the formula

$$d = \frac{t\, v_1 v_2}{2\sqrt{v_2^2\, v_1^2}}$$

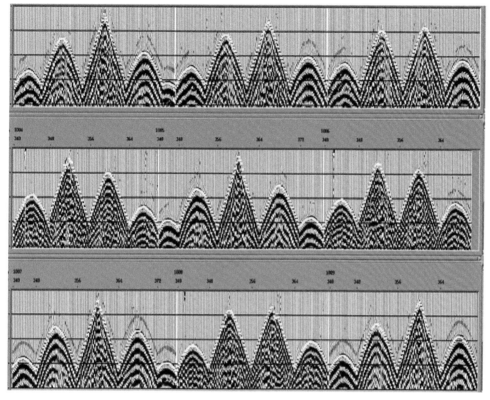

FIGURE 5.8   First break picked on all of the records. *Source: Geotrace – Geometry and Refraction Statics.*

Figure 5.9 shows the effect of refraction statics correction. Events are lining up after refraction statics is applied to the data.

FIGURE 5.9   On the left is a seismic data before static correction. On the right is the same data after static correction. *Source: Geotrace – Geometry and Refraction Statics.*

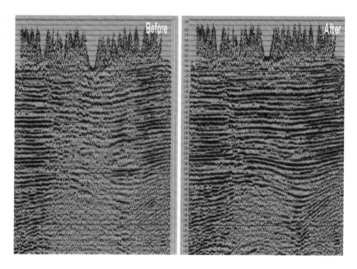

DATUM AND FLOATING DATUM

In seismic data, the term *datum* refers to an arbitrary planar surface to which corrections are made and on which sources and receivers are assumed to lie to minimize the effects of topography and near-surface zones of low velocity.

The datum moves shots and geophone groups from the actual surface to a new reference plane underlain by material having a higher velocity than the weathered layer.

By definition, the datum level is the depth level at which the shots and geophone groups would have been positioned if there is no elevation and topography effect. That is, a datum is a reference surface (Figure 5.10).

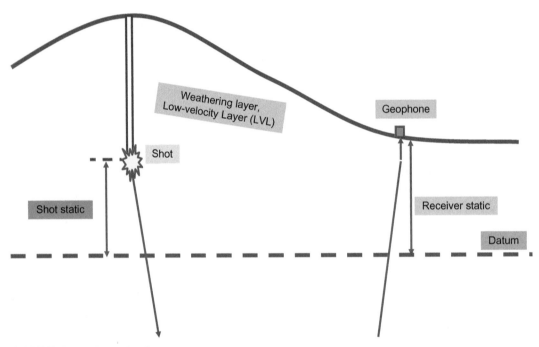

FIGURE 5.10   Datum level.

For land data, the shots and geophone groups can be at any depth; therefore, a floating datum is used which is a smooth line that follows the general elevation trends along the line, but removes any rapid shot to shot and geophone group to geophone group variation along the line.

The floating datum may be above or below the shot or geophone position and the resultant statics may be positive or negative. When the floating datum is below the shot datum, time is taken off the events, so a negative statics correction is applied. When the floating datum is above the shot datum time is added to the event, so a positive static correction is applied.

For marine data, the datum is the shot (airgun) and receiver (streamer) position, just below the sea.

NOISE ATTENUATION

In an ideal seismic survey, the recorded seismic trace from each shot will be at identical position in space and time, and only random noise will differ from shot to shot. Therefore, to attenuate noise the geophysicists add geophones output which have the same reflections point but different random noises. Figure 5.11 shows seismic data with noise (left) and the same data after noise is attenuated (right).

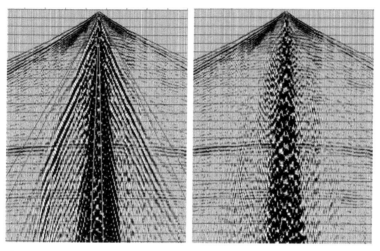

FIGURE 5.11   Seismic data before and after attenuation. *Source: arCIS SEISMIC SOLUTION.*

In Figure 5.12, for each shot the geophysicists will get a separate record and the noise is different for each record. If the geophysicist adds the record of several shots together, this will enhance the signal and reduce random noise. This process is called vertical stack or vertical summing.

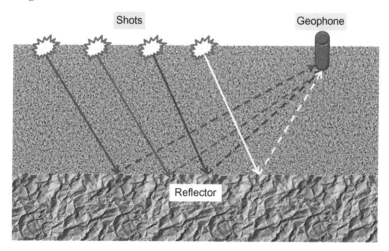

FIGURE 5.12   Several shots into a single geophone.

Note that recording several shots into a single geophone, each of the recorded traces will have the different noises. This is because the recordings are made at different times. But in the case of a single shot into different geophones (Figure 5.13), the noise in the traces is different because the geophones are in different position.

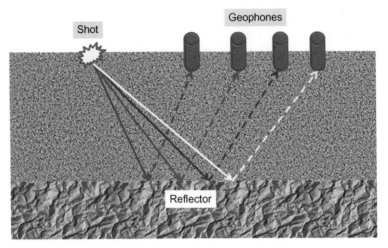

FIGURE 5.13   A single shot into several geophones.

Common Midpoint Stack

Stacking is the process whereby traces are summed to improve the signal-to-noise ratio, reduce noise and improve seismic data quality. Traces from different shot records with a common reflection point, such as common midpoint (CMP) data, are stacked to form a single trace during seismic processing. Stacking reduces the amount of data by a factor called the fold.

In CMP stack, noise is attenuated by adding all the geophones output with the same reflections' point but different random noises.

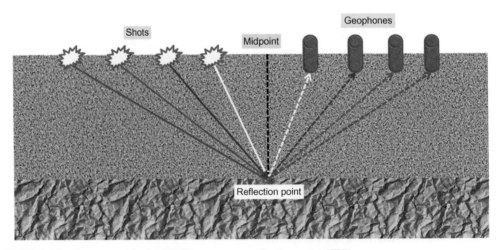

FIGURE 5.14   Common midpoint (CMP) or common reflection point (CRP).

The four shots and geophones in Figure 5.14 have the same reflection point, recorded with the noise observe at different times and at different places because the shots are taken at different times. Since the shots to geophones distances are different, the surface wave is different as well. By adding together the reflected signal obtained from one shot to the corresponding geophone, the geophysicists build up the reflections and suppressed both random noise and surface wave.

In Figure 5.14, the shots to geophones distances are different; therefore, the reflection path lengths are different as well. During seismic data processing, the reflection times are adjusted to compensate for the different path lengths.

Note that in principle, seismic reflection section is a series of seismic traces recorded by a geophone at the same location as the shot. Each trace must be time-corrected to allow for the source-geophone offset, the correction depending on the layer velocities. If the correction is accurate, a given reflection is moved up the trace to the position it would have were the source and receiver coincident. After correction, the traces are summed.

Note that noise attenuation usually precedes deconvolution. This is because the source wavelet which spiking deconvolution shortens can be clearly seen.

TRACE EDITING

Trace editing includes de-spiking (eliminating of high-amplitude anomalies), polarity reversal (change to correct polarity) and trace zeroing (set trace amplitudes to zero if average amplitude is outside amplitude thresholds).

Seismic data are examined to spot bad traces from each seismic line that contains spikes or noise trains that are unrelated to the "true" seismic data and to spot any trace with incorrect polarity.

A bad recording group will affect different traces on different shots. For example, a bad receiver will generate one noisy trace per shot, and its location with the shot record will move as the shot moves.

Land data contain noise train from the shot itself. This noise, which travels along the near surface, is usually referred to as ground roll. The traces that contain the noise train and spikes will be muted out (that is, replace with zero), as they can spread in both time and space by subsequent processing steps if not removed before hand.

Note that muting is the removal of the contribution of selected seismic traces in a stack to minimize air waves, ground roll and other early-arriving noise. Low-frequency traces and long-offset traces are typical targets for muting.

POLARITY REVERSAL

One final problem in trace editing is that of polarity. Polarity is the way in which seismic data are recorded and displayed. Most seismic data are recorded using the standard specified by the Society of Exploration Geophysicists (SEG).

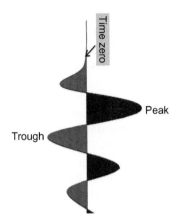

FIGURE 5.15   Recorded seismic trace.

SEG normal polarity data have compressional wave from positive reflection coefficient boundaries (Figure 5.16) recorded as a negative number on tape and displayed as a trough. A negative reflection has the same shape but is reversed – every peak a trough, every trough a peak (Figure 5.15).

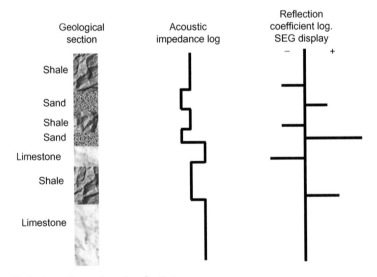

FIGURE 5.16   Geologic section and earth reflectivity.

Note that reflection coefficients are positive if acoustic impedance increases across the boundary (sand to shale) and negative if acoustic impedance decreases across the boundary (sand to shale).

Soft Kick

Soft kick is high to low impedance (shale to sand), that is, negative reflection coefficient.

SEG – displayed top of sand or base of shale as negative loop.
Non-SEG – displayed top of sand or base of shale as positive loop.

Hard Kick

Hard kick is low to high impedance (sand to shale or shale to limestone), that is, positive reflection.

SEG – displayed base of sand or top of shale as positive reflection coefficient.
Non-SEG – displayed base of sand or top of shale as negative reflection coefficient.

Note that the geophysicists need to spot those traces within the seismic record that have reverse polarity with respect to all other traces and check the above convention.

Polarity reversal of a trace simply requires the computer to multiply every sample in the trace by -1.

Note that the geophysicists only reverse the trace once in the processing sequence.

Understanding CMP Binning and Sorting

CMP BINNING

The seismic survey area (that is, the location where the data were acquired) is divided into regular spaced rectangular bins with all the midpoints in each bin cross-indexed by line, shot and receiver group number.

CMP bin is a square or a rectangular area, which contains all midpoints that correspond to the same CMP (Figure 6.1).

FIGURE 6.1   Survey lines showing 3D grid.

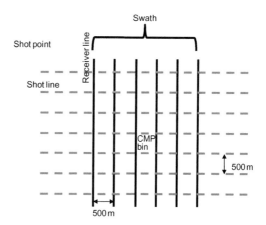

Note that to the geophysicist seismic trace lives at the midpoint between the source and receiver offset. Therefore, traces that fall into the same bin are stacked and their number corresponds to the fold of the bin (Figure 6.2).

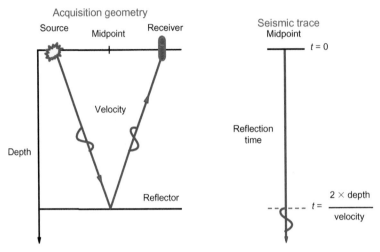

FIGURE 6.2   Conceptualized the midpoint position where seismic trace is located.

The bin size corresponds to the length and to the width of the bin. Smallest bin dimensions are equal to half source point interval and half receiver interval. For example, if the source point interval is 50 m and the receiver point interval is 50 m, the smallest bin size is 25 × 25 m.

The bin size will affect the horizontal resolution of the seismic data and its frequency content.

Resolution is defined as the ability of a seismic method to distinguish between two stratigraphy sequences in a seismic section. High frequency and short wavelengths give better vertical and lateral resolution. Note that resolution deteriorates with depth and with increasing velocities.

Vertical seismic resolution is the minimum resolvable stratigraphy thickness. The minimum stratigraphy thickness is one-quarter (1/4) of the seismic wavelength or two-way time thickness is the half of the dominant seismic period.

Horizontal (lateral) seismic resolution corresponds to the direction parallel to the seismic measurement plane. It is related to the Fresnel zone.

The Fresnel zone is defined as the subsurface area which reflects energy that arrives at the earth's surface within a time delay equal to half the dominant period ($T/2$). Note that seismic migration can improve lateral resolution by reducing the size of the Fresnel zone.

SORTING

A seismic line is made up of many shot gathers, each consisting of a number of traces. A seismic trace may contain from hundreds to thousands of individual samples, with each seismic shot generating 24 or 48 traces. A typical seismic line will contain several millions of samples. When all of the traces associated with one shot are gathered together, it is called a shot record or a field record. When several shot records have been acquired, it is possible to sort, or gather, a subset of traces from the entire data set in various ways. Therefore, we can say that shot record is the fundamental unit that makes up seismic lines (Figure 6.3).

FIGURE 6.3   3D shot point. Six panels (lines) of 240 receivers each.

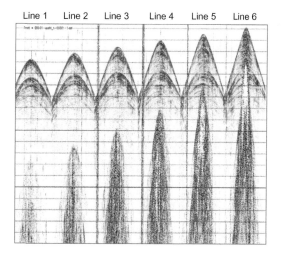

HOW DATA IN A SEISMIC LINE ARE SORTED

Common Offset Gather

Common offset gather is made up of all traces whose shot–receiver offset is the same. Common offset traces differ in receivers, sources and reflection points. In a common offset gather, trace-to-trace time differences are caused by differences in dip and/or velocity (Figure 6.4).

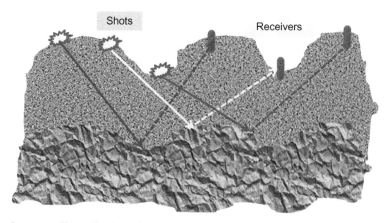

FIGURE 6.4   Common offset reflection points.

Common Receiver Gathers

Common receiver gathers are not commonly used in processing but can be useful to identify bad trace. A bad receiver will generate one noisy trace per shot, and its location with the shot record will move as the shot moves. Associating the bad traces with a particular receiver may be difficult until common receiver gathers are viewed (Figure 6.5).

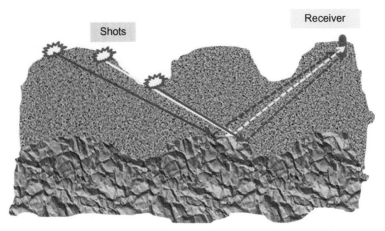

FIGURE 6.5   Reflections from different shots recorded by a receiver.

Common Midpoint Gathers

CMP gather corresponds to all traces that fall into the same midpoint and they are very important in seismic data processing. Normal move-out (NMO) is applied in CMP gathers. Common midpoint traces differ in move-out, offset, receivers and sources. CMP gathers are used as input to velocity analysis and CMP stack after NMO/DMO (Figure 6.6).

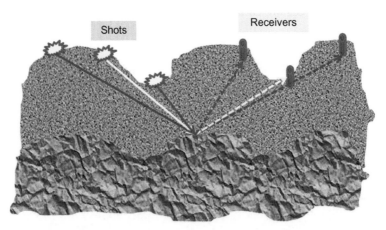

FIGURE 6.6   Common midpoint reflections.

In 3D survey, many traces are associated with each midpoint location. Since the earth consists of many reflecting interfaces, the reflection points will be distributed in 3D space and time, depending on the details of the subsurface velocity and structure.

After applying various processing technique to the data, all pre-stack traces having the same midpoint will be stacked (summed) to give a single trace that lives at that midpoint. This stack traces are an approximation to the zero offset seismic trace that would have been recorded at this location (Figure 6.7).

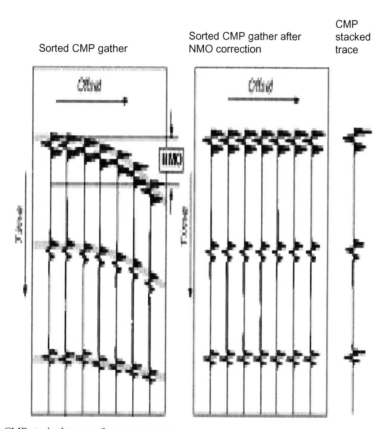

FIGURE 6.7   CMP stacked traces. *Source: www.engsoc.org.*

The stack traces are displayed together to form a stack seismic section (Figure 6.8).

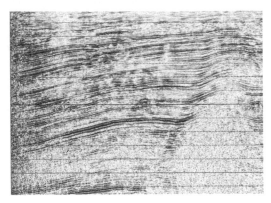

FIGURE 6.8   3D stack seismic section.

Common Shot Gathers

Common shot gathers correspond to all traces recorded from a single shot. Common shot traces differ in move-out, offset and receivers and may differ in dip (Figure 6.9).

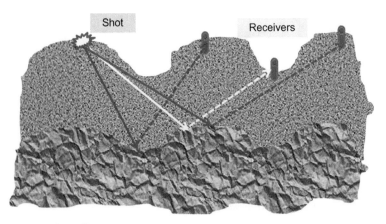

FIGURE 6.9   Common shot reflections.

Understanding Deconvolution

In seismic data acquisition, an energy source (dynamite on land and airgun in marine) is used to produce acoustic wave that propagates through the earth (Figure 7.1). The propagated acoustic wave is quickly distorted due to absorption and attenuation (attenuation is the loss of energy or amplitude of waves as they pass through rocks) that occurs in the earth. Also, the subsurface is full of multiple and ghost reflections and these make it difficult to determine which reflections are returning from which reflectors. The seismic section that resulted from this process does not represent the true image of the subsurface.

On the seismic section, you will notice that many of the horizons have the same amplitude and thickness (Figure 7.2). In other to eliminate these unwanted effects and produce a seismic section that is clear and easy to interpret, the process called deconvolution is applied to the data.

Note that as the acoustic wave (seismic signal) radiates through the subsurface, the amplitude of the seismic signal decreases due to spherical divergence and also higher frequencies within the seismic signal are attenuated due to absorption. By restoring these high frequencies, deconvolution can improve the resolution of the seismic section (Figure 7.3) by producing a broad bandwidth with high frequencies and provide the geophysicists with a more accurate image of the subsurface.

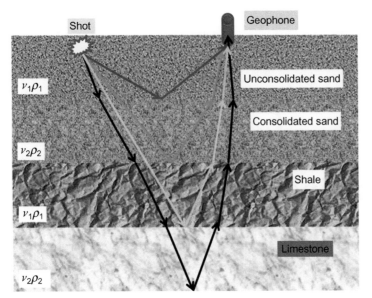

FIGURE 7.1 Conceptualized seismic acquisition.

Shot

Geophone

$v_1\rho_1$

Unconsolidated sand

Consolidated sand

$v_2\rho_2$

Shale

$v_1\rho_1$

Limestone

$v_2\rho_2$

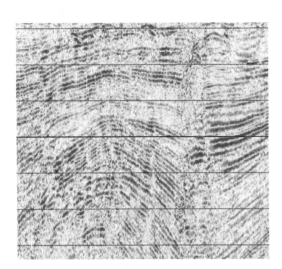

FIGURE 7.2 Seismic section before deconvolution.

FIGURE 7.3 Seismic section after deconvolution.

Note that deconvolution is usually performed as early as possible in the processing sequence to stabilize the frequency content of the data and to remove multiple reflections prior to velocity analysis.

DECONVOLUTION PROCESS

The processes of deconvolution start with the convolutional model of the seismic trace in time domain. The source wavelet (propagated acoustic wave) that is sent into the subsurface convolves with the earth reflectivity series (contrast in acoustic impedance at layer interface) to produce the recorded seismic trace. Mathematically, convolution is given by

$$S_t = R_t * W_t + \text{noise}$$

where S_t is the recorded seismic trace, R_t is the earth reflectivity, and W_t is the source wavelet (Figure 7.4).

FIGURE 7.4   Conceptualizes the convolutional model of seismic trace.

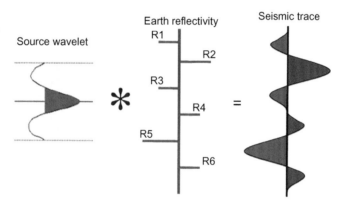

Deconvolution reverses the process used to acquire the seismic trace and begins with the recorded seismic trace.

By definition, deconvolution is the separation of the earth reflectivity and the source wavelet or the shortening of the source wavelet to a signal as short as possible. This is done by compressing the source wavelet embedded on the seismic trace into a spike so that the recorded seismic trace contains only the earth reflectivity (Figure 7.5). This will improve the resolution of finer details in the seismic data and makes reflections more visible and easy to interpret.

Note that short pulses (spike) have large bandwidth and it is easier to time the arrival time of a spike.

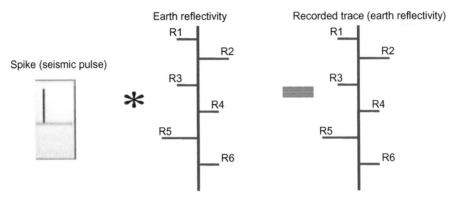

FIGURE 7.5   Conceptualized deconvolution processes.

DIFFERENT KINDS OF WAVELET GENERATED BY SEISMIC SOURCES

It is very important to know the source wavelet when performing deconvolution. Therefore, before going further let's define phase and different kinds of wavelet produced by seismic sources.

Phase is defined as the initial position of a wavelet at time zero. There are two types of wavelets that are important in seismic. These are minimum-phase and zero-phase wavelet.

MINIMUM-PHASE WAVELET

The shape of the source wavelet generated from an explosive (dynamite) on land seismic survey is a minimum-phase wavelet (Figure 7.6).

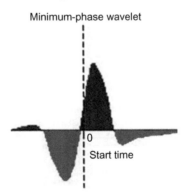

FIGURE 7.6   Minimum-phase wavelet.

A minimum-phase wavelet has no amplitude before a definite start time. The minimum-phase wavelet has the greatest amount of its energy at time zero. This does not necessarily mean that the leading loop is the largest; some minimum-phase wavelets have their greatest amplitude in the second loop.

The reflection time for a minimum-phase wavelet is measured at the start of the wavelet where the trace first breaks down or up.

ZERO-PHASE WAVELET

Vibroseis is sometimes used as the seismic source in some region on land seismic survey. The shape of the source wavelet generated by a vibroseis is a zero-phase wavelet (Figure 7.7).

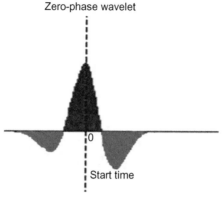

FIGURE 7.7   Zero-phase wavelet.

A zero-phase wavelet has amplitude at its start time. The wavelet is symmetrical about the zero time and has energy at negative time.

The reflection time for a zero-phase wavelet is at the central peak or trough, so picking reflection times is much easier and accurate for zero-phase wavelets. This is one reason why zero-phase wavelets are usually output in seismic data processing.

Note that in seismic, the cosine wave is the sinusoid of reference since it is a maximum at time zero and is an even or symmetrical wavelet.

In seismic data processing we generally assume that the acquired seismic data is a minimum phase and that after deconvolution the data is now zero phase.

MIXED-PHASE WAVELET

Airgun used as the seismic source in offshore seismic survey produces a mixed-phase wavelet (Figure 7.8).

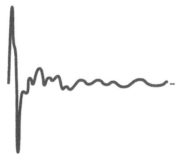

FIGURE 7.8   A mixed-phase wavelet.

HOW TO ESTIMATE THE SOURCE WAVELET USE FOR DECONVOLUTION

One of the techniques used to determine the source wavelet for deconvolution is through VSP (vertical seismic profile) survey.

VSP is a geophysical measurement made in vertical wellbore with geophones position at depth inside the wellbore and a source at the surface near the well (Figure 7.9).

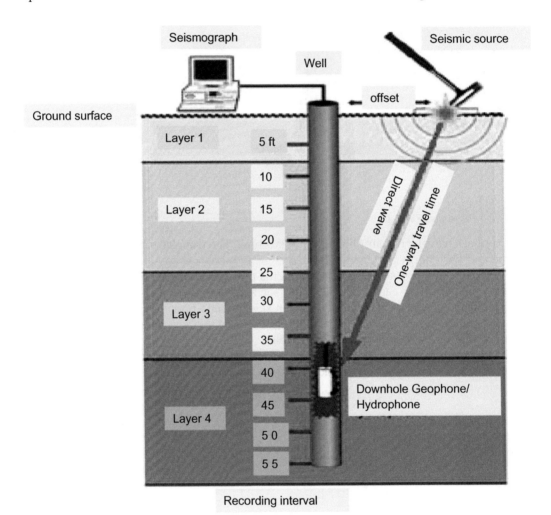

FIGURE 7.9   Schematic of vertical seismic profile. *Source: Resolution Resource International. www.rri-seismic.com.*

When the seismic source is detonated at the surface, the geophones receive the direct acoustic wave. With VSP survey, the geophysicists measure the seismic signal at depth and use this measurement to extract the source wavelet.

If Figure 7.10a is the recorded seismic trace, then a filter can be design to extract the source wavelet (Figure 7.10b) from the seismic trace.

When the source wavelet is known, it will then be possible to calculate its inverse and apply it to the seismic section to perform deconvolution.

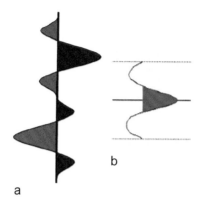

FIGURE 7.10   Recorded seismic trace (a) and source wavelet (b).

Note that if the source wavelet is reasonably consistent from shot to shot, then the inverse of one source wavelet may be computed and apply to the whole survey to perform deconvolution. That is, an inverse filter (deconvolution filter) is designed for this source wavelet to give a filter that can be applied to every single trace in the survey.

INVERSE FILTER

When the seismic signal produced from the seismic source propagates through the subsurface, the seismologists record reflections that are reflected off layered boundaries. These reflections are equal to the impulse response of the earth and the earth impulse responses include reflectivity series and some undesirable effects, such as attenuation, ghosting, and multiple reflections. In deconvolution the geophysicist estimates these effects as linear filter and then designed and apply inverse filter. The inverse filter is also called deconvolution filter.

Recall that the aim of deconvolution is to restore high frequencies that were filtered by the earth as the seismic energy propagates through the earth, attenuates multiple reflections, balances amplitude in the seismic data, and simplifies the shape of the embedded source wavelet in the seismic data so that the geophysicists can see a trace that can be interpreted as independent events. Therefore, the geophysicist has to design the time domain inverse filter that will reshape the source wavelet.

When we discussed convolution in Chapter 2, we get to understand that the auto-correlation of a seismic trace in the time domain is equivalent to multiplying the amplitude spectrum by itself and subtracting the phase spectrum from itself in the frequency domain.

Therefore, auto-correlating the input trace, cross-correlating the input and desired output, and solving the least square equations for a filter of a reasonable length, the geophysicists produced an inverse filter which can then be applied to the seismic section to perform deconvolution.

Note: Autocorrelation compares a trace to itself. Autocorrelation is useful in the identification of multiple or ghost reflections and in the designing inverse filter to suppress them. The inverse filter can improve seismic data that were adversely affected by convolution that occurs naturally as seismic energy propagates through the subsurface.

Figure 7.11 shows a section before and after applying the inverse filter (deconvolution filter). After applying inverse filter, the reflections are visible, the amplitude spectrum is broader, and the autocorrelation is more centre lobed. You can also notice that the frequency content has improved.

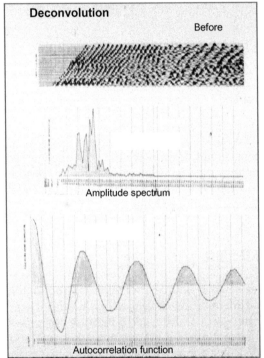 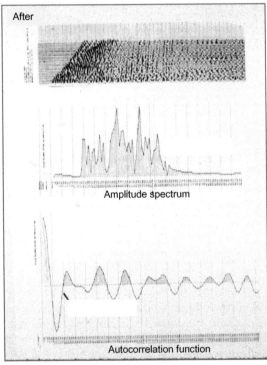

FIGURE 7.11   A synthetic data, amplitude spectrum, the autocrrelation function before and after applying deconvolution. *Source: Reflection seismic data processing lecture slide after Steve H. Danbom, Ph.D., P.G. (Spring 2007).*

Note: In the frequency domain, seismic trace is expressed as amplitude and phase as a function of frequency. This is referred to as amplitude and phase spectrum.

PARAMETERS THAT CONTROL DECONVOLUTION

Four parameters are very important in any deconvolution methods. They are as follows:

- The filter length

Firstly is the length of the inverse filter. As the length of the inverse filter increases, the deconvolution result improves. However, if the inverse filter is too long, the very information the geophysicist is trying to extract may be distorted.

Also, because one of the purposes of deconvolution is to remove short-period multiples from the recorded seismic data, it is therefore important that the filter length, specified in milliseconds, is long enough to include at least two bounces of the maximum reverberation time of the multiple reflections the geophysicist is trying to remove. This is because a strong multiple generators will generate multiples for both the down-going wave and the up-going wave. The multiple of the down-going wave is removed by a filter, which has the same length as the multiple periods (Figure 7.12).

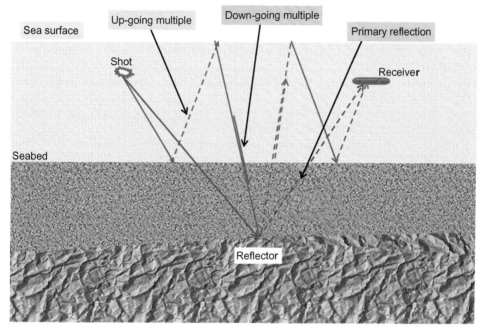

FIGURE 7.12   Up-going and down-going multiple reflections.

- The gap length

    A gap, in milliseconds or samples, is inserted into the filter to prevent the filter from changing the wavelet close to every reflector.

    If the objective is to shorten the wavelet, the gap should be zero. A gap length within the range of 0–1 implies spiking deconvolution. Higher gap implies predictive deconvolution. A longer gap is used when amplitude is to be preserved (for AVO processing).
- The design gate

    A "window" needs to be specified, that is, the length of the seismic trace that would be auto-correlated. This is called the design gate. It is important that the gate is long in other to obtain a good sampling of the subsurface geology. The design gate is usually limited to the part of the seismic data that contains the best signal.

- White noise

The fourth critical parameter is the percentage of white noise or pre-whitens that the geophysicist adds to the input data. By applying pre-whiten the effect cause by non-random noise can be lessen.

Note that in cases where the deconvolution equation cannot be solved the value that needs to be added for it to be solved is usually of the order of zero point one (0.1) to one percent (1%), and it is usually referred to as the white noise level. This is done in the frequency domain.

TYPES OF DECONVOLUTION

Spiking Deconvolution

In spiking deconvolution, it is assumed that the auto-correlation of the seismic trace is the same as the auto-correlation of the source wavelet, and that the reflectivity is random and its auto-correlation is a spike.

In other words, the desired output is specified as a spike, and the entire trace is converted into a single spike. This has the effect of equalizing the amplitude spectrum within the trace.

Figure 7.13 shows the effect of spiking deconvolution on a common offset data. After applying spiking deconvolution, the frequency content and resolution of the data have improved significantly. Event can be seen clearly on the seismic section.

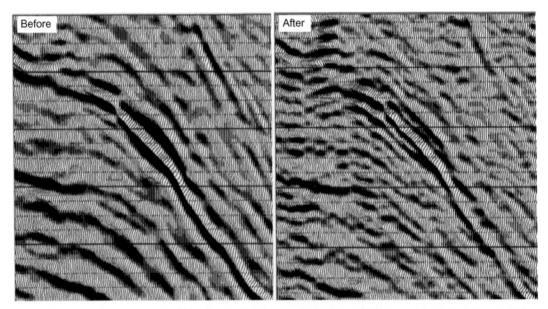

FIGURE 7.13 Seismic section before spiking deconvolution and the same data after applying spiking deconvolution. *Source: The Leading Edge. intl-tle.geoscienceworld.org.*

Predictive Deconvolution

In predictive deconvolution, the predictable component of the seismic trace which is the multiple reflections is eliminated, while the reflectivity series which is the unpredictable component of the seismic trace is left untouched.

Predictive deconvolution removes short-path multiple reflections.

Figure 7.14 shows the effect of predictive deconvolution. Looking at the shot gather after applying predictive deconvolution, you will notice that reflection continuity has improved and events are easier to trace.

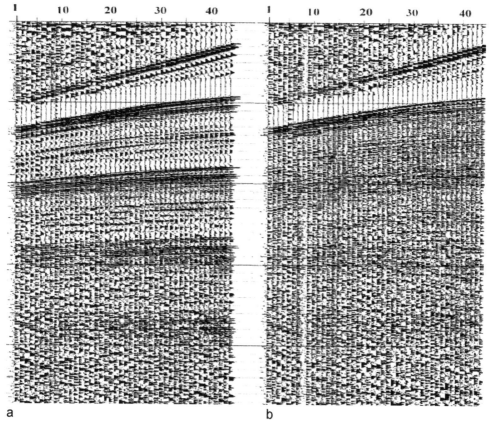

FIGURE 7.14   (a) Shot gather with short-period multiples. (b) The same shot gather after application of predictive deconvolution to eliminate the multiples. *Source:* ELSEVIER Journal, *Volume 160, issue 3–4, 1 September 1999.*

Understanding Sample Data

SAMPLE DATA

Seismic field data are recorded digitally as a sampled data at regular time interval. In other to reshape the seismic trace the geophysicists need to know the sample value which was use to record the data. Before going further let's define some terms:

Sample interval: Sample interval is the time or distance between data points.

Sample rate: Sample rate is the number of measurements per unit of time.

Sampling frequency: Sampling frequency is the number of sample (data) points in a unit time (Figure 8.1).

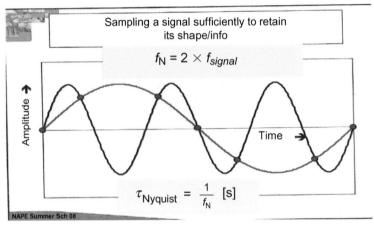

FIGURE 8.1   Shows sample points (red dot). *Source: NAPE Summer Sch 08.*

SAMPLE PERIODS/RATE

Below are the various sample periods used to sample seismic trace:

One millisecond (1 ms)
Two milliseconds (2 ms)
Four milliseconds (4 ms)
Eight milliseconds (8 ms)

FREQUENCIES CONTENT OF SEISMIC DATA AT VARIOUS SAMPLE PERIODS

The frequencies content of seismic data at various sampling periods are:

- For one millisecond (1 ms) sampling period; recall that frequency, f is given as:

$$f = \frac{1}{T}$$

But $1\,\text{s} = 1000\,\text{ms}$

$$f = \frac{1}{1 \times 10^{-3}}$$

$$f = \frac{1000}{1} = 1000\,\text{Hz}$$

But

$$f_N = \frac{1}{2}f$$

where f is the frequency of the input data and f_N is the Nyquist frequency

$$f_N = \frac{1000}{2} = 500\,\text{Hz}$$

Therefore, for 1 ms sampling period the highest frequency in the seismic data is 500 Hz.

- For two milliseconds (2 ms) sampling period, the frequency content of the seismic data is:

$$f = \frac{1}{T}$$

$$f = \frac{1}{2 \times 10^{-3}}$$

$$f = \frac{1000}{2} = 500\,\text{Hz}$$

$$f_N = \frac{1}{2}f$$

where f is the frequency of the input seismic data and f_N is the Nyquist frequency:

$$f_N = \frac{500}{2} = 250\ \text{Hz}$$

Therefore, for 2 ms sampling period the highest frequency in the seismic data is 250 Hz.

- For four milliseconds (4 ms) sampling period, the frequency content of the seismic data is:

$$f = \frac{1}{T}$$

$$f = \frac{1}{4 \times 10^{-3}}$$

$$f = \frac{1000}{4} = 250\ \text{Hz}$$

$$f_N = \frac{1}{2}f$$

where f is the frequency of the input seismic data and f_N is the Nyquist frequency

$$f_N = \frac{250}{2} = 125\,\text{Hz}$$

Therefore, for 4 ms sampling period the highest frequency in the seismic data is 125 Hz.

Using the same method above, the Nyquist frequency for eight milliseconds (8 ms) sampling period is

$$f_N = \frac{125}{2} = 62.5 \text{ Hz}$$

Therefore, for 8 ms sampling period the highest frequency in the seismic data is 62.5 Hz.

ALIASING

Aliasing is the distortion of frequency introduced by inadequately sampling the seismic data, which results in ambiguity between the seismic data and noise.

NYQUIST FREQUENCY

The maximum frequency beyond which aliasing will occur is called the Nyquist frequency. Above Nyquist frequency, seismic signal with higher frequencies are reconstructed in the form of seismic signal with lower frequencies.

The highest frequency that can be successfully recovered from sampled seismic data is one half divided by the sample period. That is,

$$f_N = \frac{1}{2\,dt}$$

where dt is the sample rate or period and f_N is the Nyquist frequency. Note that if the sampled seismic data contains frequency components greater than $1/2\,\Delta t$, aliasing will occur. To avoid this, an anti-alias filter is usually applied. Table 8.1 gives a summary of Nyquist Frequencies for various sample periods.

TABLE 8.1   Nyquist Frequencies for Various Sample Periods

Sample period (ms)	Nyquist frequency (Hz)
1	500
2	250
4	125
8	62.5

Note that the farther the data is sampled below the Nyquist frequency, the less worry there is of distorting amplitudes which is important to AVO and reservoir property prediction.

Seismic data can be acquired, processed and interpreted at different sample rate or period. Whenever the time sample rate is double, the size of the seismic data is cut into half. Note that the data file that is stored in the computer is not the size of the survey area, but the size of the sampled data.

Note also that the lower the sample rates the more the sample points and more expensive to process the acquired seismic field data.

RESAMPLED SEISMIC DATA

Resampling is the process of reducing seismic data to a low sample rate or higher sample interval. If there are no frequencies in the original data above the new Nyquist frequency, then the process is reversible.

Why Seismic Data Are Resampled

There are two main reasons why seismic data are resampled:

* The reason why seismic data are resampled is because there is rarely good information in the acquired seismic data at frequencies greater than 100 Hz at depth of interest because the bandwidth of the seismic source is usually limited.
* And also, the earth attenuates the higher frequency content of the propagating acoustic wave (source wavelet) that is sent into the earth at a greater rate than the lower frequencies. This leads to a progressively lower frequency wavelet and poorer seismic resolution with depth.

SEISMIC BANDWIDTH

Seismic data are bandlimited and the bandwidth of most seismic data that contain useful information is between 5 and 100 Hz. Bandwidth affects the resolution of seismic data. Wide bandwidth is very important because the wider the bandwidth of the seismic data the better it is to image the subsurface.

Figure 8.2 shows the frequency range (bandwidth) that can be seen in this seismic data.

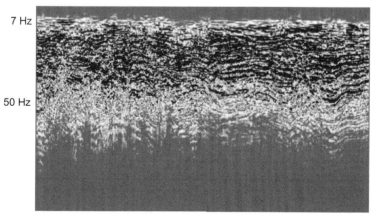

FIGURE 8.2   Seismic data frequency spectra and the bandwidth that can be seen in the data.

WHEN ARE THE ACQUIRED SEISMIC DATA RESAMPLED IN THE PROCESSING SEQUENCE?

Seismic data are resampled at any stage in the data processing sequence after applying de-convolution. Note that de-convolution is applied at the same sample rate as the acquired field date.

Seismic data, acquired at 2 ms sample period, correspond to a frequency of 500 Hz, with a Nyquist frequency of 250 Hz, will normally have frequencies that can also be handled by a larger sample rate (4 ms). Therefore, the data are resampled to 4 ms sample rate to keep the high frequencies because they improve the resolution of the data.

Note that seismic data cannot be resampled to a sample rate with a Nyquist frequency lower than the highest usable frequency in the data. For example, if the maximum seismic frequency in the data is 90 Hz, a time sample rate of 4 ms is appropriate to resample the data. Although 8 ms sample rate might be used to resample the data to reduce the data volume, the higher frequencies in the data will be aliased and the geophysicists need the high frequencies because they improve the resolution of the data.

Example 1

If we resample a 40 Hz seismic data with a sample period of 2, 4 and 8 ms, what is the frequency of the seismic data after resampling?

Answer

The frequency content of the (that is, output frequency) seismic data after resample at 2, 4 and 8 ms will be the same as the original seismic data because 40 Hz is below the Nyquist frequency of the three sampled period.

Example 2

If we resample a 90 Hz seismic data with a sample period of 2, 4 and 8 ms, what is the frequency of the seismic data after resampling?

Answer

The frequency of the seismic data resample at 2 and 4 ms will be the same as the original seismic data because 90 Hz is below the Nyquist frequencies of both sampled period. However, the frequency of the seismic data sampled at 8 ms will be different from the original seismic data of 90 Hz because 90 Hz is above the Nyquist frequency for 8 ms sampled period.

The Nyquist frequency for 8 ms sampling period is 62.5 Hz, which means that the data to be sampled is 27.5 Hz above the Nyquist frequency. Aliasing will occur and the seismic signal after resample will be 35 Hz.

Example 3

If we resample a 140 Hz seismic data with a sampling period of 2, 4 and 8 ms, what is the frequency of the seismic data after resampling?

Answer

The frequency content of the seismic data resample at 2 ms will be the same as the original seismic data because 140 Hz is below the Nyquist frequency (250 Hz) for 2 ms sample period.

However, the 140 Hz original seismic data is 15 Hz above the Nyquist frequency (125 Hz) for 4 ms sample period. Therefore, aliasing will occur and the frequency content of the seismic data at 4 ms after resample is 110 Hz.

Also, the 140 Hz seismic signal is 77.5 Hz above the Nyquist frequency for 8 ms sample period, which is 62.5 Hz. Aliasing will occur. Therefore, the frequency content of the seismic data at 8 ms after resample is 15 Hz.

Note that to prevent aliasing, a filter must be applied before sampling or resampling to a large sample period, and a filter must be applied to limit frequencies below Nyquist.

FILTERING

Filtering usually means frequency filtering. So, how do we use the frequency domain to filter or resample seismic data? To use the frequency domain to filter seismic data, the geophysicists simply do the following:

Step 1

Apply Fourier transform to transform the seismic signal (trace) from time domain to frequency domain.

Step 2

Apply a filter to remove the high frequencies above Nyquist frequency.

Applying a filter means multiplying the filter amplitude spectrum with the trace amplitude spectrum and adding the filter phase spectrum to the trace phase spectrum.

At frequencies where the filter amplitude response is less than 1, attenuation takes place. Therefore, the idea of filtering is also to reduce or attenuate frequency components were noise dominate over our seismic signal (trace).

Step 3

Transform the seismic trace back to time domain using the inverse Fourier transform.

LET'S DEFINE SOME IMPORTANT CONCEPT RELATING TO FILTERING

Filters are defined by parameters called cut-off frequency and attenuation rate. The cut-off frequency is the frequency at which the response is −3 dB.

Attenuation Rate

The attenuation rate is usually stated as the number of dB/octave the filter amplitude response decreases beyond the cut-off frequency.

Low-Cut-Off Filter

A low-cut-off (high pass) filter applied to a seismic trace attenuate frequencies less than the cut-off frequency.

High-Cut-Off Filter

A high-cut-off (low pass) filter applied to a seismic trace attenuate frequencies above the cut-off to make the output more low frequency. The high-cut-off filter is very important as it prevent aliasing.

Band Pass Filter

A band pass filter rejects frequencies below its low-cut-off frequency and above its high-cut-off frequency while passes a range of frequencies within the acceptable limits of the filter unaltered.

Band Rejects Filter

A band rejects filter attenuate frequencies above the limits of a filter.

SPATIAL RESAMPLING

When seismic data are resampled in the space dimension it is called spatial resampling.

In Figure 8.3, as the seismic shot is detonated each geophone (receiver) within a group records a different part of the reflected seismic signal. The signal (trace) arrives later at the geophones, which are the furthest away from the shot. The smaller the group interval, the higher the frequency the reflected seismic trace can contain without aliasing. This is why the highest frequency required to properly image an exploration objective is an important parameter when designing data acquisition programs.

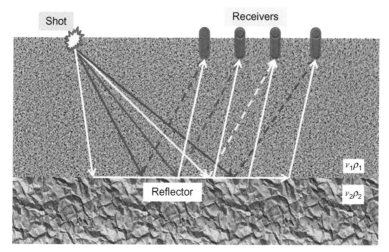

FIGURE 8.3   Seismic reflection and how the signals are recorded.

Note that wave number is the reciprocal of wavelength and wave number is sometimes called spatial frequency.

Wave number $(k) = 1/$wavelength (λ).

Trace Mixing

One way to spatially resample seismic data is by combining pairs of traces from each shot and so reducing the number of traces per shot by a factor of 2, or by reducing the number of traces in the receiver panel by combing shots.

Trace mixing can be performed after applying time-shifts to each of the traces. This slant-stack will bias the mix towards enhancing certain 'dips' in the data and is sometimes known as beam-steering. Note that, any trace mixing carry out during seismic processing, introduces some form of spatial filtering on the data.

Spatial Filtering

Spatial frequencies can be analyzed by Fourier transform. The term frequencies means frequencies in time or Hertz, and the term spatial frequencies or wave number means frequencies in space. All recording system, which uses arrays of geophone or hydrophones, introduces spatial filtering in the recording signal (trace).

Figure 8.4 is an array of geophones (receivers) spaced about, let say 3 m apart. The geophones are arranged 'in-line' with the shot, so that the reflections will be arriving from either the left or the right. These reflections will arrive at a slightly different time at each geophone. This means that when the traces from all the receivers are summed together, some form of spatial filtering will be introduced to the combined trace.

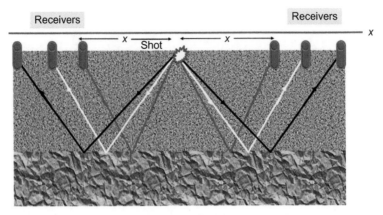

FIGURE 8.4   This set-up describing spatial filtering.

FK Analysis

When the shot in Figure 8.4 is fired, the reflected signals are recorded at the receivers (geo-phones) positions which are located at discrete point. So, in seismic, sampling is done both in time and space.

Since the seismic trace are recorded in time, Fourier transform is use to transform the recorded trace from time domain to frequency domain. Fourier transform can also be used to transform seismic trace in space (x) to wave number (k) also called spatial frequency.

Seismic velocity is expressed as

$$v = f\lambda$$

But, wavelength (λ), is

$$\lambda = \frac{v}{f}$$

where f is the frequency.

The reciprocal of wavelength (λ) is wave number (k).

That is, wave number (k) = 1/wavelength (λ)

$$k = \frac{1}{\lambda}$$

Therefore,

$$k = \frac{f}{v}$$

Note that because seismic data are sampled in time and space, data outside a plane, that is, between $-k$ and $+k$ in the wave number direction and between zero (0) and f in the frequency direction is called the FK plane (Figure 8.5).

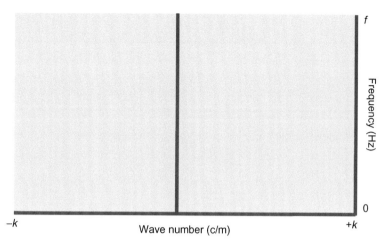

FIGURE 8.5   Shows FK plane.

k is wave number or spatial frequency, f is frequency and v is velocity. FK analysis allows geophysicists to examine the frequencies (f) and spatial frequencies or wave number (k) at the same time.

To transform a trace in time/space into 'F' and 'K', each trace is Fourier transformed from time to frequency, and then every frequency component is Fourier transformed in space to give an FK spectrum.

Use of FK Analysis

FK domain is used for noise attenuation. Noise such as ground roll, direct rays and first break refraction can be separated from the primary reflection in the FK domain.

The FK domain provides the geophysicist with an alternative measure of the content of seismic data in which operations such as filtering of certain unwanted events can be done more effectively.

Note: When seismic data are transformed into the FK domain, dipping events in time/ space become dipping events in frequency/wave number.

The Tau–P transform transforms dips into single points in a time/dip domain so that the data can be edited in the 'dip' domain and transform back into the time/space domain.

Understanding NMO, Velocity Analysis DMO, and Stacking

The seismic processes discussed so far improve the signal of each separate trace. Before the traces are stack (to improve the signal-to-noise ratio), the trace must be time-corrected to allow for the source–geophone offset and the correction depends on the layer velocities.

NORMAL MOVE-OUT

A seismic reflection section, in principle, consists of a series of traces recorded by a geophone (receiver) at the same location as the shot. In this case, the ray incident on a particular reflector at right angles would be reflected along the same travel path towards the receiver; the zero-offset seismic section is a normal incidence section. The reflection would be recorded at normal incidence two-way travel time (Figure 9.1).

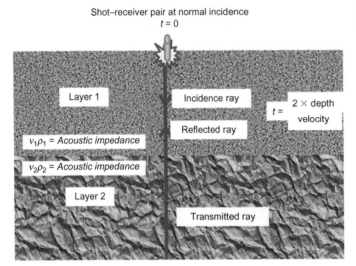

FIGURE 9.1 Zero-offset seismic reflection.

Practically, the shot and receiver are always at some distance (offset) from one another (Figure 9.2). This will result in distortion in reflections (hyperbolic curve in Figure 9.3), which are delayed due to the increase in travel time of the reflected ray with longer offsets.

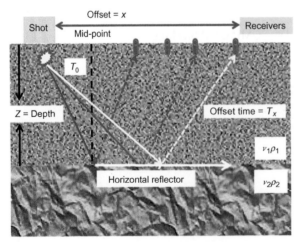

FIGURE 9.2 Reflection at non-normal incident.

FIGURE 9.3 An hyperbolic curve due to increase offset.
Source: tle.geoscienceworld.org.

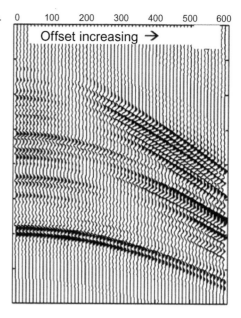

HOW TO DETERMINE THE MAGNITUDE OF CORRECTION REQUIRED TO FLATTEN THE HYPERBOLIC EVENT

The vertical distance from the mid-point between the shot and the receiver to the horizontal reflector in Figure 9.2 is

$$Z = \frac{T_0 V}{2}$$

where Z is depth, V is velocity, and T_0 is zero offset time or normal incidence two-way time.
    The distance travelled along the ray path from reflection point to receiver due to the offset is

$$Z_X = \frac{T_X V}{2}$$

where T_X is the two-way time for the offset (x) and V is the velocity.
    For a horizontal reflector (Figure 9.2), the reflection travel–time curve (hyperbola) for different offsets (x) between the source and the receiver is calculated using

$$T_X = \sqrt{\frac{X^2}{V^2} + T_0^2}$$

In the presence of dipping reflector (Figure 9.4), the reflection travel-time curve for different offsets (x) between the source and the receiver is calculated using

$$T_X = \sqrt{T_0^2 + \frac{X^2 \cos\theta^2}{V^2}}$$

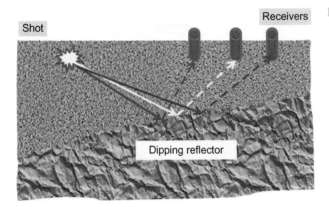

FIGURE 9.4   Dipping reflector.

By plotting the offset travel time (T_X) as a function of offset (x), we get

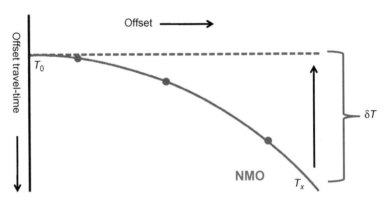

FIGURE 9.5   Travel-time curve for different offsets.

T_0 is the true zero-offset time (Figure 9.5). Note that the variation in reflection arrival time due to variation in source and receiver offset appears as a hyperbola on a CMP gather (Figure 9.6a).

To eliminate the effect of offset, the geophysicists move each trace at time T_X up to time T_0 in Figure 9.5, so that each reflector is horizontal in the CMP gather (Figure 9.6b). This correction is known as normal move-out correction (NMO correction) and it is the difference between the offset time (T_X) and the zero-offset time (T_0).

$$\Delta t_n(x) = t(x) - t_0 = \sqrt{t_0^2 + \frac{x^2}{V^2}} - t_0$$

The NMO correction, $\Delta t_n(x)$, is the time-shift necessary to convert the observed time at any offset to the zero-offset time.

Note that the parameters defining the hyperbolic NMO correction are offset (x), normal incidence travel time (t_0) and velocity V.

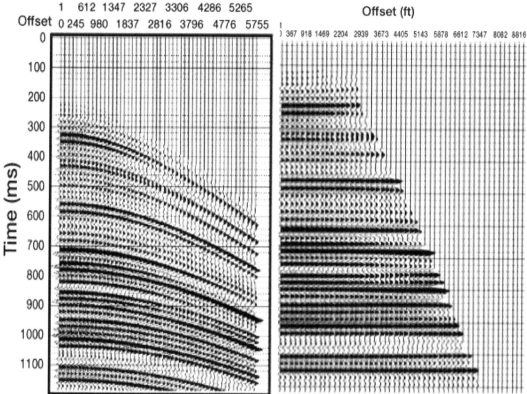

FIGURE 9.6   (a) Reflections hyperbola on the CMP gather and (b) the same CMP gather after NMO corrected.

NMO CORRECTION

The NMO correction is defined as the time-shift that is applied to each seismic trace within a CMP gather in order to make all the traces from a particular reflector equal in time. The correction depends on the layer velocities.

DYNAMIC CORRECTION

Dynamic correction is a time-variant operation performed on a seismic trace.

Looking at Figure 9.7, you will notice that the hyperbolic curves reduce down the record (that is, flattens with depth). The reason is that the difference between the zero-offset (T_0) and the far-offset time (T_X) decreases for deeper reflections. The NMO correction changes with time and this is referred to as dynamic correction. The dynamic correction therefore corrects for the hyperbolic distortion of events known as normal move-out (NMO).

Offset (m)

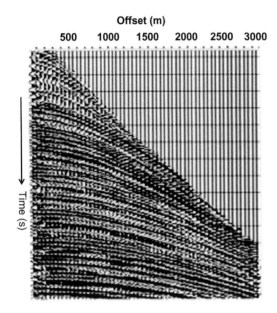

FIGURE 9.7   Hyperbola on a CMP gather decreases with time (that is, becomes flat with depth).

NMO STRETCH

After applying NMO correction on the CMP gather in Figure 9.8, to correct for normal move-out, you will notice that the NMO curves overlap at shallow data. This is because the computer shifted the far-offset traces more than the near-offset traces to the correct time. This is known as NMO stretch.

The NMO process introduces wavelet changes, particularly at far offsets. As the traces are stretched, its frequency content is compressed. That is, the NMO has the effect of stretching the wavelet, which compresses the bandwidth. This is an undesirable effect because it reduces seismic resolution. The severely stretched portions will be muted (that is, replace with zero) through a technique called stretch mute.

Note that the application of muting does not cause a problem as the far offsets and shallow times contain ground roll, direct arrival, refracted and mode-converted energy which should also be muted out.

DIFFERENT KINDS OF SEISMIC VELOCITIES

Before we go into details in explaining velocity analysis, let's define some kinds of seismic velocities.

Average Seismic Velocity

The average seismic velocity is the distance travelled by a seismic wave from the source to a reflector in the sub-surface and back to the surface. That is, average seismic velocity is two-way distance divided by travel time and is given as

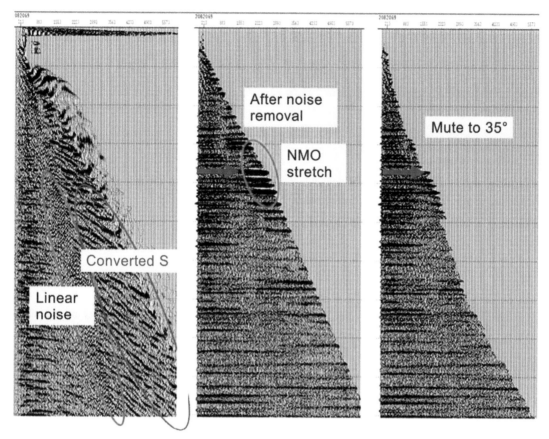

FIGURE 9.8   (a) A CMP gather processed through an NMO correction. (b) An NMO stretch area. (c) The stretch area has been muted out. *Source: tle.geoscienceworld.org.*

$$v_a = \frac{2z}{2t} = \frac{2z}{T}$$

where z is depth to reflector and T is two-way time (Figure 9.9).

Interval Velocity

Interval velocity is defined as the thickness of a stratigraphic layer divided by the time it takes to travel from the top of the layer to its base:

$$v_i = \frac{2\Delta z}{2\Delta t} = \frac{2\Delta z}{\Delta T}$$

Note that the layer thickness is equal to the isopach value of the interval.

The average velocity can be calculated from the interval velocities. The equation for the average velocity in terms of interval velocity is (Figure 9.10)

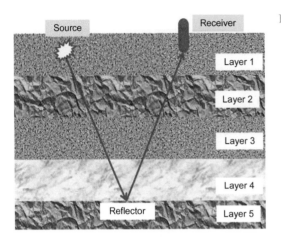

FIGURE 9.9   Described average velocity.

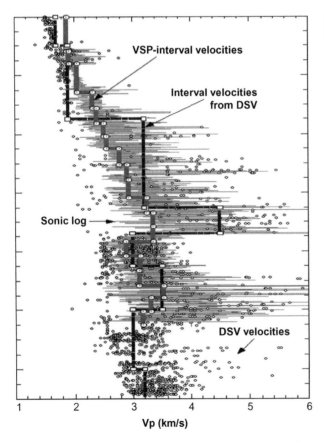

FIGURE 9.10   Sonic log and interval velocities. *Source: www.ideo.columbia.edu.*

$$v_a = \frac{\sum_{i=1}^{n} v_i \Delta t_i}{\sum_{i=1}^{n} \Delta t_i}$$

Note that for layer along the surface of the earth, the average velocity equals the interval velocity. But this is not true for deeper layers.

Instantaneous Velocity

Instantaneous velocity is the velocity at which a seismic wave propagates at a point within the sub-surface. The closet to instantaneous velocity measurement is the sonic log.

Stacking Velocity

The velocities derived from seismic data are stacking velocities. The velocities are determined using velocity analysis.

Stacking velocity is the velocity that gives the optimum CMP or CDP stack output when used for NMO corrections. Note that it is not the true rock velocity.

At shallow depth, the stacking velocities go close to rock velocities. As depth increases, the more imprecise the stacking velocities are when compared with rock velocities.

Note that for horizontal reflectors, the velocity calculated from the move-out hyperbolas does not correspond to the average velocity. The velocity (stacking velocities) calculated from move-out hyperbolas are always faster than the average velocity.

NMO Velocity

The NMO velocity is the velocity used to correct for NMO – to make reflections time on CMP gather occurs at the same time on all traces.

Difference Between Stacking and NMO Velocities

Stacking and NMO velocities differ on the basis of spread length.

The stacking velocity is basically a parameter that generates the best stack. It is based on the travel-time hyperbola that best fits data over the entire spread length.

NMO velocity is based on the small-spread hyperbolic travel time.

RMS Velocity

The root-mean square (RMS) velocity is the value of the square root of the sum of the squares of the stacking velocity values divided by the number of values. The RMS velocity is that of a wave through sub-surface layers of different interval velocities along a specific ray path. RMS velocity is higher than the average velocity. RMS velocity is calculated using

$$v_{RMS} = \sqrt{\frac{\sum_{i=1}^{n} v_i^2 \Delta t_i}{\sum_{i=1}^{n} \Delta t_i}}$$

For a horizontal layer or gently dipping layer, NMO and stacking velocities are equal to the RMS velocity. However, stacking velocities differ substantially from the RMS velocities in areas with large lateral variations in velocity.

THE DIX FORMULA

The Dix formula is an equation used to calculate the interval velocity between any two points on a time velocity graph established from NMO hyperbolas (Figure 9.11):

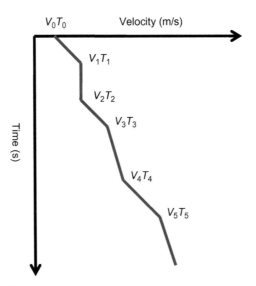

FIGURE 9.11   Velocity function.

$$V_i = \sqrt{\frac{V_2^2 T_2 - V_1^2 T_1}{T_2 - T_1}}$$

where V_1 and V_2 are the velocities at times T_1 and T_2, respectively, and V_i is the interval velocity.

If the layers are flat and the offset is small, Dix equation gives good estimates of interval velocities using stacking velocities. Dix equation is not valid for steeply dipping layers and complex geology.

VELOCITY ANALYSIS

Velocity analysis is the process whereby the velocity that flattens the reflection hyperbola on a CMP/CDP gather is determined. The velocity is called the stacking velocity.

In velocity analysis, velocities are picked at selected locations along the seismic section and the results are linearly interpolated from one analysis location to the other as long as the picks are from the same reflector and the same reflection time.

For example, if the velocity picked at 2 s in Figure 9.12 is 2000 m/s and the velocity picked at 2 s at another location is 2500 m/s, then the interpolated velocity function is 2250 m/s. If a change in depth occurs between the velocity samplings, the interpolated velocity function is wrong and that causes the NMO correction at that location to be wrong.

FIGURE 9.12   Describes velocity analysis.

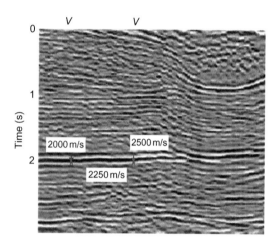

Note that the velocities picked at various CMP or CDP locations of the seismic line during velocity analysis are called velocity function.

Because rock velocities generally increase with depth due to higher pressure and more compaction, the velocity function, that is the function of measured velocity as a function of travel time, also increases.

Velocity Analysis Techniques

The techniques used to determine the velocities required to flatten the hyperbola on a CMP gather are discussed below.

Constant Velocity Gathers (CGV)

A constant velocity gather display (Figure 9.13) shows a single CMP gather displayed with different velocities to allow the best velocity for NMO correction to be chosen by observing the velocity at various times down the record that results in the flattening of the reflection hyperbolas in each event.

Figure 9.13 shows a CMP gather from a seismic line, processed through NMO correction with a series of constant velocities to establishing by eye the best velocity at each time.

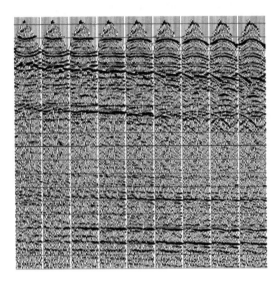

FIGURE 9.13   A single CMP gather displayed with different velocities to allow the best velocity for an NMO correction to be chosen. *Source: Shell International Exploration and Production B.V.*

Constant Velocity Stack (CVS)

This is carried out for several CMP gathers and the NMO-corrected data are stacked and displayed as a panel for each different stacking velocity. Stacking velocities are picked directly from the constant velocity stack panel by choosing the velocity that yields the best stack response at a selected event (Figure 9.14).

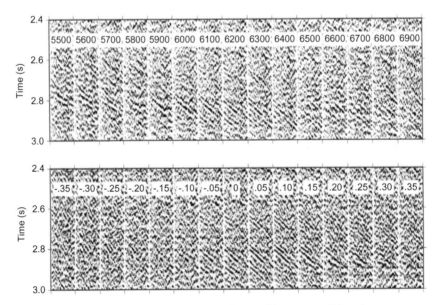

FIGURE 9.14   A constant velocity stack panel. *Source: Johiris Rodriguez, www.123people.se.*

Semblance

A semblance is a quantitative measure of the coherence of seismic data from multiple channels that is equal to the energy of a stacked trace divided by the energy of all the traces that make up the stack. If data from all channels show continuity from trace to trace, the semblance has a value of unity.

In other words, the semblance of a small window of stacked data is effectively a measure of how well the stack compares with the individual components going into the stack.

Note that stacking is the sum of the number of samples in a trace divided by the number of traces. For example, if there are five time samples each from three traces, the stack is their sum divided by 3 (Table 9.1). The semblance value equals the sum of the amplitudes squared after stack divided by the average sum of the amplitudes squared before stack. The semblance value will always be between zero (which indicates no match) and one (which means a perfect match). If the computed semblance is plotted against effective velocity (Figure 9.15a), together with interval velocity (Figure 9.15b) and a stack section (Figure 9.15c), the geophysicists will have a very powerful tool for analyzing velocities.

TABLE 9.1   Five Time Samples from Three Traces

	Trace 1	Trace 2	Trace 3	Stack
Time	−150	−120	48	−74
	990	−375	645	420
	−647	1799	249	467
	540	−432	1677	595
	−359	−50	139	−90

The stack is their sum divided by 3.

What the geophysicist needs to do is to pick the semblance peaks that provide the best stack at a series of times down the record. Although this is a rapid process, it may be repeated at 500-m intervals throughout a 1000-km survey.

POTENTIAL PROBLEMS IN VELOCITY PICKING

Recall that the objective of velocity analysis is to determine the stacking velocity or the seismic velocities of layers in the sub-surface. And in doing so, there are some potential problems associated with velocity picking.

Firstly, multiple reflections appear on a velocity analysis as a fake semblance peaks among the real primary reflections. So how can the geophysicists distinguish these multiples from the primary reflections?

Primary Reflections

Primary reflections are generated by energy which is reflected only once at a single reflector.

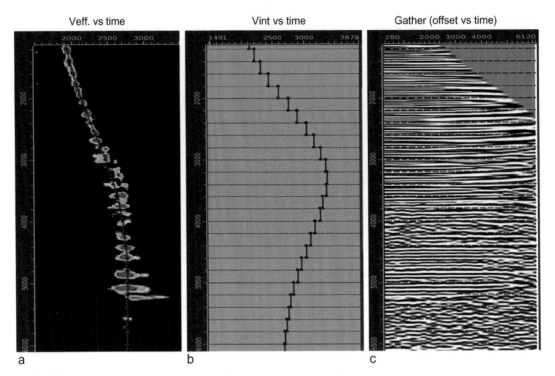

FIGURE 9.15   Velocity analysis using semblance. *Source: www.liag-hannover.de.*

Multiple Reflections

Multiple reflections are events which reflect more than once in any layer in the sub-surface or at the seabed and the air–water interface, as in the case of marine seismic data (Figure 9.16). The reflected wave associated with multiple reflections will have spent more time travelling through the swallower parts of the section than the wave creating the primary.

Depending on the time delay of the multiples from the primary reflections, multiples are characterized as short or long multiple reflections. Short-period multiple reflections appear as a copy of the primary reflections, while long-period multiples appear as separate events.

Generally, rock velocities tend to increase with depth. So, the multiple reflections are expected to be slower than the primary reflections at the same reflection time. For this reason, the higher range of velocities are normally picked on a semblance display.

It is important to note that velocity inversion in shallow section results in multiple reflections having velocities that are faster than the primary reflections.

Secondly, multiple reflections always have more move-out, that is, hyperbolic curvature than primary reflections.

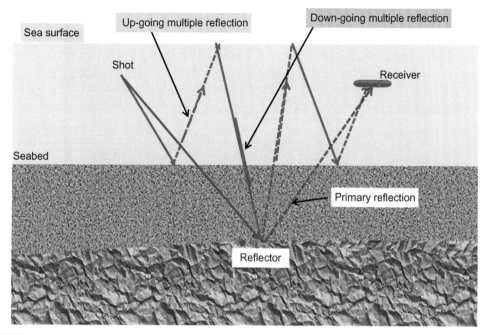

FIGURE 9.16   The figures shows how multiple reflections are formed.

Note that if the wrong velocities are picked as the primary event and all the traces are added together to form the stack (from CMP), some of the primary reflections will be attenuated.

QUALITY CONTROL OF PICKED VELOCITIES

Having picked the velocities (stacking velocities), the picked velocities are usually quality check for their spatial consistency. One of the common methods of spatial checking is by examining the iso-velocity contours along the seismic line.

Iso-velocity displays are colour coding that shows times at which the velocity functions have the same value. They are used to evaluate velocity picking (Figure 9.17).

The axes of the iso-velocity display above are space and time. The colours represent the interpolated value of the velocities function (picked velocities) at discrete intervals throughout the seismic section. The blue represents low velocities and the orange high velocities. A gross error in velocities will usually show up fairly easily on an iso-velocity display.

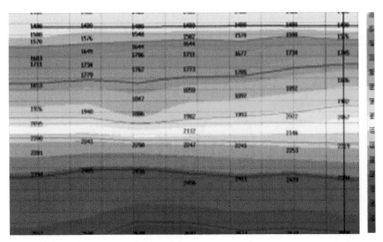

FIGURE 9.17   Iso-velocity contour. *Source: Khan, K.A., Akhter, G., 2011. Work flow shown to develop useful seismic velocity models. Oil Gas J.*

VELOCITY FIELD

The picked seismic velocities from velocity analysis are used to build a 3D volume of the velocity field, which can then be examined in any direction. So when carrying out a velocity analysis of a three-dimensional survey, the three-dimensional volume of the velocities constructed (colour display in Figure 9.18) is known as a velocity field.

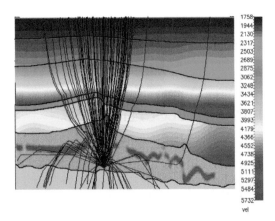

FIGURE 9.18   Ray tracing for a single reflector through the earth, modelled by a 3D velocity field display in colour. *Source: DEVELOPER ZONE and Bernard Deschizeaux CGGVeritas. http.developer.nvidia.com.*

DIP MOVE-OUT

The velocities have been picked, and the necessary dynamic corrections applied to correct the data to zero-offset. But the geophysicists still don't have a common depth points (CDP) data, and the recorded data are clearly not below the mid-point in the case of a dipping reflector.

The assumption that common mid-point (CMP) implies common depth point (CDP) may be true for horizontal reflector. But if there are dipping events within the seismic section, the seismic data will not be recorded at the correct spatial position, and the assumption that a common mid-point (CMP) is the same as a common depth point (CDP) breaks down.

Figure 9.19 shows a common mid-point or common reflection point for a horizontal reflector. The black-dotted line shows a normal incidence ray at right angle to the reflector, and it indicates the mid-point between the shots and the receivers. The blue, brown, and green lines show reflections for three different offsets.

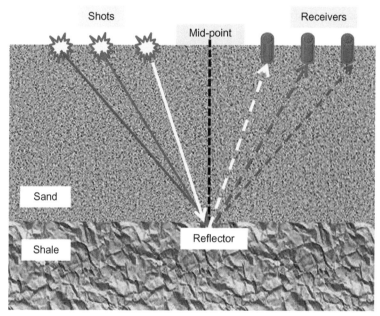

FIGURE 9.19   Horizontal reflector.

If the reflector is a dipping interface (Figure 9.20), the reflection points move 'up-dip', away from the nominal 'CMP' in the centre of the diagram and the actual points being recorded are different for different offsets. The apparent velocity obtained from the CMP gathers is also incorrect.

The apparent velocity is the velocity seismic wave appears to propagate along the surface of the earth. For events with constant dip, the velocity will be wrong by a factor equal to 1 over the cosine of the dip.

If the lithology has a constant velocity of 1800 m/s, the apparent velocity will be, say, 1950 m/s. Note that as the dipping angle increases the reflection point moves 'up-dip' and the apparent velocity increases.

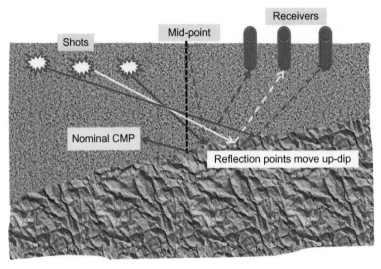

FIGURE 9.20   Dipping reflector.

In the presence of dipping event, adding traces of gather means adding reflections from extended area of the reflector rather than a nominal point. This will smear the trace and possibly decrease resolution. In other to stack the traces correctly, the geophysicists need to correct all the traces in the reflector to its true zero-offset position and ensure that velocities are no longer dependent on dip.

Dip move-out (DMO) technique is applied to reposition the data so that there are one common reflection points for all the traces in the gather in the individual offset planes. DMO move the data received at a particular time into all possible zero-offset positions in each offset 'plane'. In doing this for every time and CMP position in the common offset plane, signal is kept and noise and mis-positioned data are cancelled out.

The mathematical expression for DMO is given as

$$\Delta T = \frac{X^2 \sin^2 \emptyset}{V^2}$$

where X is the trace offset, $\emptyset$ is the angle of dip for the sub-surface interface, and V is the velocity of the overlying layer.

Note that the DMO correction is directly proportional to the trace offset and the angle of dip but inversely proportional to the velocity, and as velocity generally increases with depth, the deeper the CMP, the smaller the correction.

This section below shows seismic display before and after DMO with each section having its own set of velocities. The differences are not enormous, but you can see the random noise reduction and the enhancement of some of the fine detail in the section after DMO because of stabilization of velocities and the correction from CMP to CDP (Figure 9.21).

Before DMO   After DMO

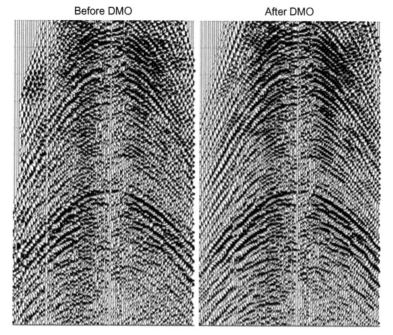

FIGURE 9.21   Seismic display before DMO (left) and after DMO is applied to the data (right). *Source: inter-geo.org.*

Note that DMO correction carried out before stack correct the mismatch between CDP and CMP while seismic migration carried out after stack, correctly position the data in space and time.

SUPER GATHERING

Velocities are normally analyzed twice before stacking the data, that is, before and after DMO (dip move-out).

Super gathering is a method of improving signal-to-noise ratio of seismic data by examining several common depth point (CDP) gathers at once during velocity analysis.

MUTE

Before the data are stacked, NMO stretch on the CDP gather and first-break noise present on the longer offsets will be removed by applying a mute to eliminate the NMO stretch and the noise.

The way it is done is that the geophysicists pick a mute and then determine the offset time in which first-break refraction and NMO stretch deteriorate the data and where the mute should start. The mute zeroes everything above the chosen time. Note that when traces are muted, it simply means that they are replaced with zeroes. This is shown in Figure 9.22.

Mute can be applied before carrying out velocity analysis but picking the mute after NMO is better because it removes any additional noise introduced by stretching in NMO.

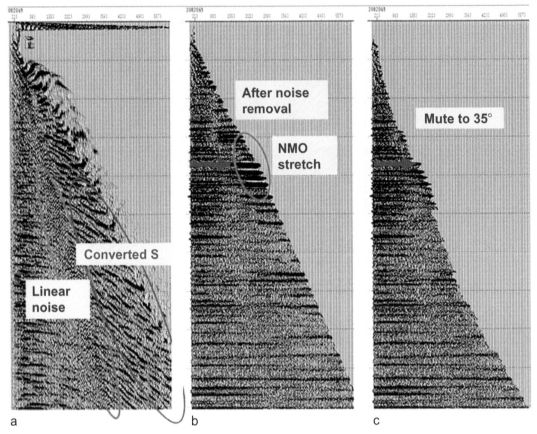

FIGURE 9.22   (a) A CMP gather processed through an NMO correction. (b) An NMO stretch area. (c) The stretch area has been muted out. *Source: tle.geoscienceworld.org.*

CMP/CDP STACKING

Each trace in the CDP has been time-corrected and move to its true common depth point or common reflection point, but the data are still not in the correct spatial position.

The geophysicists can now sum together all of the traces from this CDP into one final stack. A seismic trace may contain from 120 or more individual samples. To maintain the relative amplitude of the stack data, it must be divided by the number of traces to be stacked. The number of traces that have been added together during stacking is called the fold.

For example, if there are five traces to be stack, each of these traces have different amounts of lateral offset. When all the traces are added together, we say that is a five-fold data.

Note that each individual trace in Figure 9.23 has geologic information about the common reflection point (CDP) and certain amount of noise. The receivers at the surface pick up the reflected seismic trace and the noise. On onshore seismic survey, it might be noise from passing car/truck, recording instrument, wind, etc. On offshore seismic survey, it might be waves, noise on the boat, or acquisition ship. The geologic information from the common reflection point is the same on all the traces, but the noise is random. When stacked, all traces carry the same signal, which is enhanced as it is in phase. However, all traces have different random noises which are out of phase and therefore cancel out.

The stacking process decreases or attenuates random noise. After stacking, random noise is attenuated by a factor equal to the square root of the fold of stack ($\sqrt{f}$). Thus, the fold contributes greatly to the enhancement of the signal-to-noise (S/N) ratio.

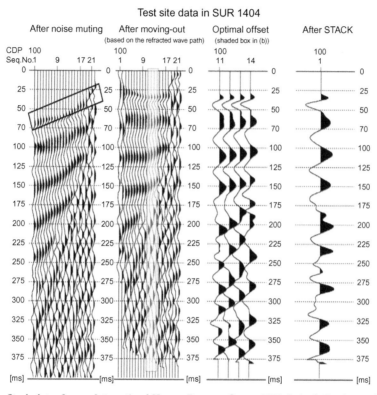

FIGURE 9.23   Stack data. *Source: International Human Resource Center, 2008. J. Appl. Geophys. 66(1–2), 1–14.*

DIFFERENT STACKING METHODS

There are different techniques used to stack seismic data.

Weighted Stack

It can be used for attenuation of multiple. It removes any AVO effects.

Median and Min-Max Stack

Median and min-max stack eliminates the highest and/or lowest amplitudes from the stack before summing. This technique may be useful to eliminate spikes and noise burst.

10

Understanding Multiple Reflections

Multiple reflections occur whenever the energy from the shot reflects more than once in any layer. Layers with strong acoustic impedance contrast can cause multiple reflections to occur. Water layers also generate multiple reflections. That is, in marine, the air–water interface and the water–bottom interface allow a sequence of strong reverberations to occur. These (multiples) appear with regularity at successively later times. These multiples obscure the primary reflections and make interpretation of the final seismic section very difficult, though primary reflections and multiple reflections appear as distinct events on a seismic section.

Figure 10.1 shows the response of water layer to a seismic source. The energy repeatedly bounces between the seabed and the sea surface.

This example demonstrates a classic multiple attenuation challenge; a complex, high-amplitude, scattered multiple wave field obscures the reservoir zone. The images in Figure 10.2 show just how effective the 3D GSMP technique by *WesternGeco* can be in removing complex multiples.

CHARACTERISTICS OF MULTIPLE REFLECTIONS

- Multiple reflections go on and on repeating with the same time interval and gradually decreasing in amplitude.

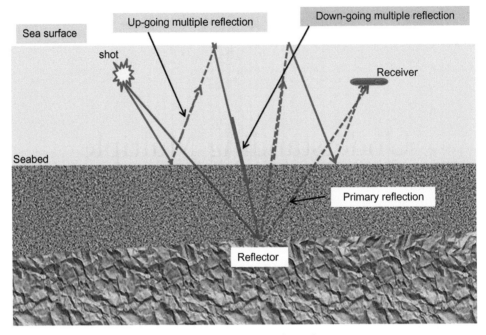

FIGURE 10.1   This figure shows how multiple reflections are formed.

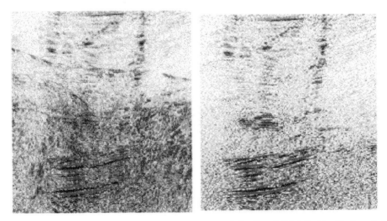

FIGURE 10.2   Seismic data before the application of General Surface Multiple Prediction (left) and after the application of General Surface Multiple Prediction (right). *Source: WesternGeco, True – Azimuth 3D General Surface Multiple Prediction. www.slb.com.*

- Multiple reflections appear on CDP gathers with a velocity slower than the primary reflections. That is, multiple reflections have slower velocities than primary reflections and therefore take longer time to reach long offset and will have large move-out values that appear as more curvature.

PROBLEMS CAUSE BY MULTIPLES

- Short-period multiples will modify the frequency spectrum of the data, adding unwanted information to the fine detail in each primary reflection. Deconvolution is used to remove short-period multiple reflections.
  Figure 10.3 shows a CMP gather before and after multiple attenuations.

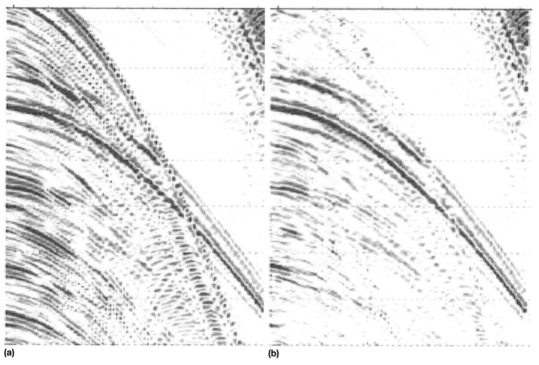

(a)   (b)

FIGURE 10.3   (a) CDP gathers with multiple reflections. (b) The same gathers after multiples attenuation. *Source: ChoiceGeophysical. www.chiocegeophysical.com.*

- Long-period multiples generate copies of the primary events at later times in the seismic record. Velocity discrimination is used to remove long-period multiples from seismic data because their travel time path is much longer than primary reflections. And also multiple reflections have slower velocity than primary reflections.

DE-MULTIPLE TECHNIQUE

Figure 10.4 is a CMP gather with strong multiple reflections before and after NMO correction, using the picked primary velocity function to correct for NMO. (a) Before NMO correction, both primary and multiple reflections have hyperbolic move-out curves. (b) After NMO correction was applied with the velocity of the primary reflections, the multiples are all under-corrected while the primary events are flattens. Note that if we apply the multiple velocity function to correct a CMP gather, the primary reflection will be overcorrected.

The multiple reflections always have more move-out, that is, curvature than the primary reflections. It is this differential move-out between the primary reflections and multiples that help in removing multiples.

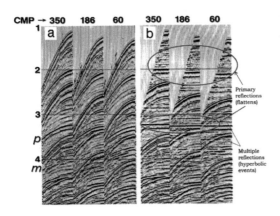

FIGURE 10.4   CMP gather before NMO correction (a) and (b) the same gather after applying NMO correction. *Source: A historical reflection on reflections – The Leading Egde by Bill Bragoset – 2005, WesternGeco, Houston, USA.*

When the traces in the gather are stacked, the primary reflections will add together coherently. Because the multiple reflections still have move-out, they will not stack coherently. Therefore, the success of CMP stacking in attenuating multiples depends on the degree of difference between the velocities of the primaries and multiples.

FK DE-MULTIPLE

Rather than applying an NMO correction with primary velocities, the gather can be corrected by using a velocity function that lies between the primary and multiple velocity functions. This has the effect of overcorrecting the primaries and under-correcting the multiples. In Figure 10.5, the events dipping upwards towards the right after applying NMO correction are the primary reflections and those dipping downwards are the multiple reflections.

Events from the CMP gather dipping downwards (multiple reflections) appear on the right-hand side on FK domain (Figure 10.6), as they have positive dip, and events from the gather dipping upwards (primary reflections) appear on the left-hand side on the FK domain (Figure 10.6), as they have negative dip. The difference in dip is used to separate primary reflections from multiple reflections. An FK filter is used to remove the positive dips in the FK domain, and these are the multiple reflections.

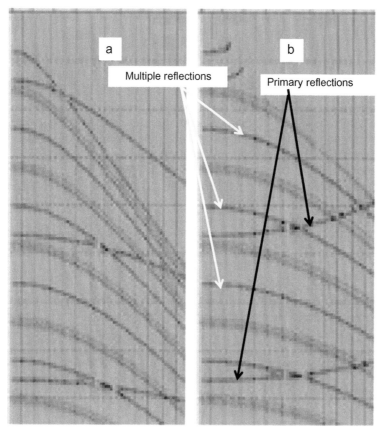

FIGURE 10.5   (a) Synthetic gather. (b) The gather with NMO corrected with velocity function that lies between the primary and multiple velocity function. *Source: ChoiceGeophysical. www.chiocegeophysical.com.*

FIGURE 10.6   FK domain. *Source: Choice-Geophysics.*

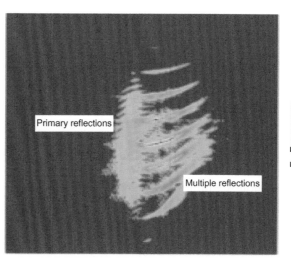

Note that FK multiple elimination will affect both primary reflections and multiple reflections at the near offset. Therefore, FK de-multiple is not appropriate for true AVO processing.

PARABOLIC TRANSFORM DE-MULTIPLE

Parabolic transform de-multiple can be used to attenuate relatively short-period multiples. This allows discrimination based on the curvature rather than the dip. Parabolic transform gives better separation of primary and multiple reflections.

Figure 10.7 shows a CMP gather and its parabolic transform. The transform summed data along constant parabolas. Primary and multiple reflections are well separated in the parabolic domain.

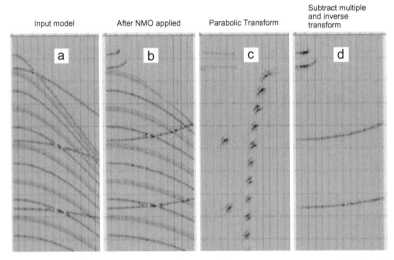

FIGURE 10.7   Parabolic transform technique used to eliminate multiple reflections. (a) CMP gather. (b) The same data after applying NMO correction. (c) The parabolic transform of the data and you can notice the separation between the primary and multiple reflections. (d) The CMP gather with multiple eliminated. *Source: ChoiceGeophysical.*

To avoid problems with any non-parabolic data, which appears as noise, the primary reflections are removed in the parabolic domain, transform back the multiples, and then subtract them from the original record. This multiple elimination technique is as good as perfect.

Parabolic de-multiple is an expensive but a very effective multiple elimination technique. Also the parabolic de-multiple preserves amplitude versus offset (AVO) effects very well.

On the stack section in Figure 10.8, you can see the improvement in the data quality after multiple attenuation.

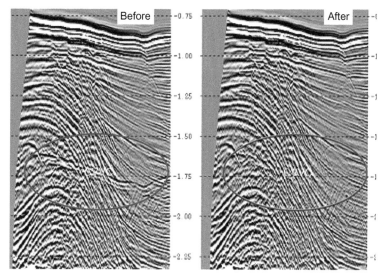

FIGURE 10.8 Stack section before multiple attenuation (left) and the same data after multiple attenuation (right). *Source: Paradigm – Echos Seismic Processing Software. www.pdgm.com.*

Understanding Residual Statics

Sometimes, after both NMO and static corrections are applied, reflections on a CMP or CDP traces are still not exactly aligned in time. This is usually due to small errors in static corrections called residual statics. Residual statics are characterized by trace-to-trace time differences that are random in nature.

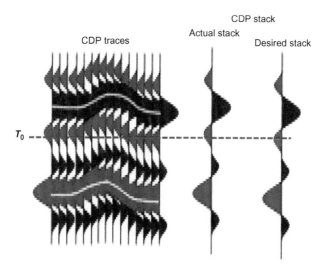

FIGURE 11.1   Conceptualized residual statics. *Source: agilegeoscience.com.*

In Figure 11.1, even though NMO corrections have been applied, the reflections do not line up across the CDP traces. When the traces are stacked two errors occurs:

Firstly, there is an error in T_o (true zero offset time).
Secondly, the amplitude and frequency are lowered.

So how do the geophysicists determine residual statics? The geophysicists use the seismic data itself to determine these residual statics.

In 3D data acquisition, the number of traces recorded from one shot, or into one receiver positions are very high. The geophysicists can make use of this data redundancy by assuming that all traces recorded from one shot will have a consistent shot static (Figure 11.2), and also all trace recorded into the same receiver group will have the same receiver statics (Figure 11.3).

This assumption is valid if all ray paths near surface are vertical, irrespective of source–receiver offset, which can be expected as near surface velocities are very low.

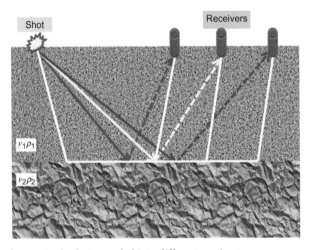

FIGURE 11.2   Traces from a single shot recorded into different receivers.

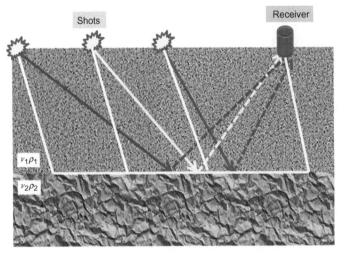

FIGURE 11.3   Traces from different shots recorded into the same receiver.

RESIDUAL STATICS CORRECTION

The technique for establishing the surface-consistent static values for sources and receivers depends on building a pilot trace section. The geophysicist finds travel times differences between shot and receivers, and between receiver and multiple shots. Because all travel time difference measurements are relative, the geophysicist, therefore, can derive these travel time differences by comparing individual traces with a pilot trace.

HOW TO BUILD A PILOT TRACE SECTION

One or more reflections are picked across the window in Figure 11.4, and the data around the picked event will be used to build a pilot trace section.

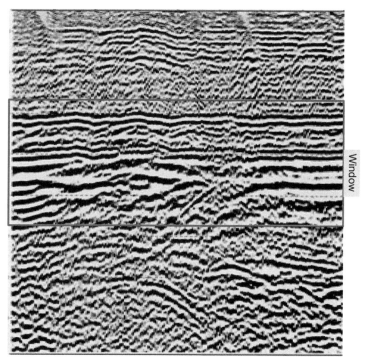

FIGURE 11.4   Seismic section before residual statics. *Source: Land Ocean Energy Services Co., Ltd. – Seismic Processing Software. www.idocean.com.cn.*

Note that the window is the length of the seismic section that would be used to build the pilot trace. A large number of adjacent traces in the window are usually smash together, usually 20 or more, aligned in time on the event. The result is a smooth section.

The geophysicists will then go back to the data before stack and compare the individual traces for every CDP against the pilot traces to establish a set of surface-consistent residual statics.

All the traces going into the stack that come from the same shot are compared with the pilot traces corresponding to the individual shot traces. The same thing is done for all traces going into the stack that comes from the same receiver.

Recall that correlation is used to measure the similarity between two seismic traces and their time difference. Figure 11.5 are two seismic traces, with a slightly time shift. This technique (correlation) is used to compare the individual input traces against the ideal pilot traces and also measure time differences by picking and then trying to rationalize these as one value for each shot and for each receiver position. If there were no residual statics present, the correlation peaks would line up on the central timing line.

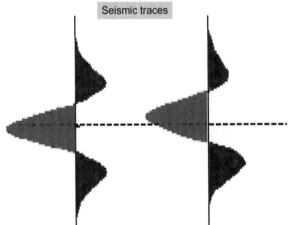

Seismic traces

FIGURE 11.5   Described trace correlation.

Figure 11.6 shows the original stacked section before and after residual statics was applied. The frequency content of the seismic data is improved and much better after residual statics was applied, and the events are more continuous.

Note that statics in the data causes complications when trying to determine dynamic corrections in order to pick velocities. Similarly, errors in velocities will complicate the determination of statics. In practice, this may mean going through cycles of residual statics and then velocity picking until the result reflects what the geophysicists are looking for.

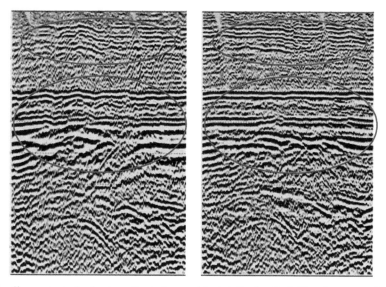

FIGURE 11.6 Shows stacked seismic section before residual statics (on the left) and the same data after residual statics was applied (on the right). *Source: Land Ocean Energy Services Co., Ltd. – Seismic Data Processing Software. www. idocean.com.cn.*

In seismic exploration, an energy source is used to send acoustic wave down the earth and part of the wave is reflected off a layer interface as shown in Figure 12.1, and the reflected wave is recorded at the surface by a receiver (geophone). When the reflector is horizontal, the reflection point lies at the midpoint between the source–receiver offset.

The two-way travel time, T, for the reflected event is twice the total distance travelled (twice the depth, Z) divided by the velocity of the layer

$$T = \frac{2Z}{V}$$

But if the reflector is dipping, then the reflection point is not directly below the midpoint between the source–receiver offset but the reflection point is located up-dip of the source–receiver offset (Figure 12.2).

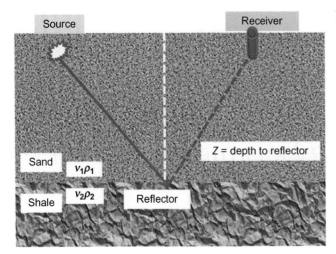

FIGURE 12.1   Horizontal reflector.

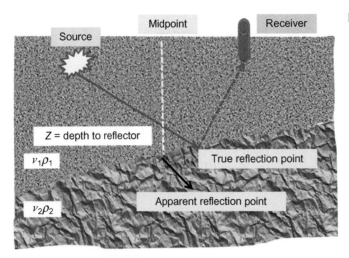

FIGURE 12.2   Dipping reflector.

The greater the dip the further the reflection point is from the midpoint between the source–receiver offset. The two-way travel time, T, of the reflected event for dipping reflector is equal to twice the depth, Z, multiply by the cosine of the dip angle divided by the velocity of the layer

$$T = \frac{2Z \cos \varnothing^\circ}{V}$$

Note that the travel time of the seismic event from a dipping interface is less than that for a horizontal interface by a factor equal to the cosine of the dipping angle.

In the case for a dipping reflector, if the seismic energy travels 0.3 s down to the true reflection point and 0.3 s up to the receiver – a total of 0.6 s. When these data are displayed on the seismic section, the reflection will be plotted 0.6 s beneath the source–receiver (apparent reflection point, A) midpoint but not at the true reflection point (that is, up-dip of the source–receiver offset).

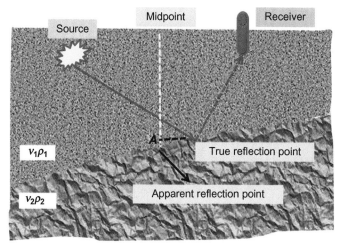

FIGURE 12.3   Apparent reflection point (A).

Therefore, when NMO and DMO corrections are applied to place all sources and receivers at the midpoint position, the reflection time is known but not their true reflection points. Thus, reflections from dipping events are plotted on the unmigrated seismic section in the wrong place. They need to be move 'up-dip' along a hyperbolic curve in order to put them in their right reflection point and the shape of this hyperbola depends on the velocity field.

Migration is used to reposition the data (Figure 12.3) from their apparent reflection point to their true reflection point in both space and time. The spatial shift use to reposition the data is calculated using:

$$\Delta X = \frac{-\text{Time} \times \text{Dip} \times \text{Velocity}^2}{4}$$

ΔX can be calculated by picking dips on the unmigrated seismic section and using the stacking velocities.

Time is in seconds (s). Dip is in seconds per metre (s/m). Velocity is in metre per second (m/s).

$$\text{Time}_{\text{new}} = \sqrt{\text{Time}^2 - \frac{4 \times \Delta X^2}{\text{Velocity}^2}}$$

Time is also calculated by picking dips on the unmigrated seismic section and using the stacking velocities derived from CMP/CDP gather using velocity analysis.

From these equations, note that any event will move up-dip after migration and deeper in time.

If we take the two migration equations, we can see that the dip term is already eliminated in the second one; we can thus derive all possible positions for the migrated event given the time of the event on the unmigrated section and given the velocity. Figure 12.4 shows unmigrated

Unmigrated image

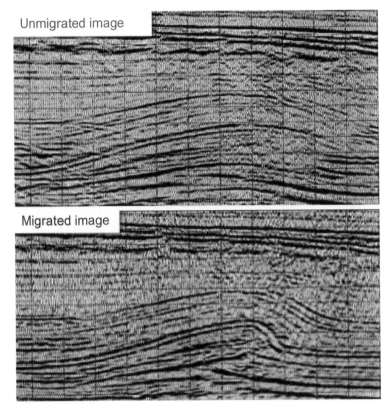

Migrated image

FIGURE 12.4 Unmigrated and migrated seismic section. *Source: AAPG Slide Resource – The Seismic Method by Fred Schroeder. Courtesy of ExxonMobil.*

and migrated seismic image. Looking at both sections you will notice a remarkable difference on the migrated image.

MIGRATION VELOCITY

Migration velocity is defined as the velocity that optimizes the repositioning of the reflected energy to the correct locations. Stacking velocities are normally use as a basis for building a 'velocity field' for migration. Figure 12.5 shows a velocity field.

Note: when carrying out a velocity analysis of a three dimensional survey, the three-dimensional volume of the velocities constructed (color display in figure 12.5) is known as a velocity field.

Note that if the data are migrated with a velocity that is too low, the data will be under migrated and if the data are migrated with a velocity that is too high the data will be over migrated.

This section was processed with the 'correct' velocities. On the unmigrated section in Figure 12.6, the resolution is poor because the reflection points are wrongly positioned.

On the migrated section, there is improvement in resolution because the reflected images are rightly position (Figure 12.7).

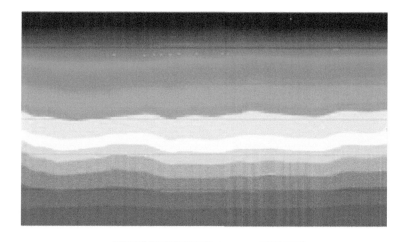

FIGURE 12.5 Velocity field. *Source: Land Ocean Energy Services Co., Ltd. – Seismic Data Processing Software.*

FIGURE 12.6 Unmigrated seismic section. *Source: AAPG Slide Resource – The Seismic Method by Fred Schroeder. Courtesy of ExxonMobil.*

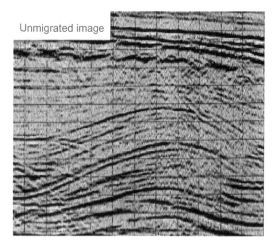

FIGURE 12.7 Migrated section of the same data as in Figure 12.8 using the correct migration velocity. *Source: AAPG Slide Resource – The Seismic Method by Fred Schroeder. Courtesy of ExxonMobil.*

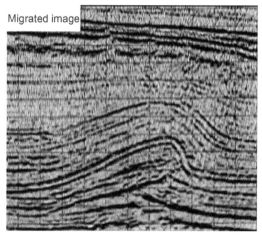

UNDERSTANDING GEOLOGIC STRUCTURES ON SEISMIC SECTIONS

Below are the appearances of some simple geological structures on unmigrated and migrated seismic section.

An Anticline

Anticlines appear wider on the unmigrated seismic sections beyond its true position and crosses through the reflections point. The higher the dips, the wider the structure will appear on the unmigrated section (Figure 12.8). On a migrated seismic section, an anticline appears compressed. If the velocities use to migrate the data are incorrect, then the final migrated structure may be narrower or wider than the true structure. This could lead to the wrong estimation of any oil or gas reserves underneath this anticline. Also, incorrect migration velocities may lead to lateral and vertical uncertainties when using the seismic data to do well-to-seismic tie during seismic interpretation.

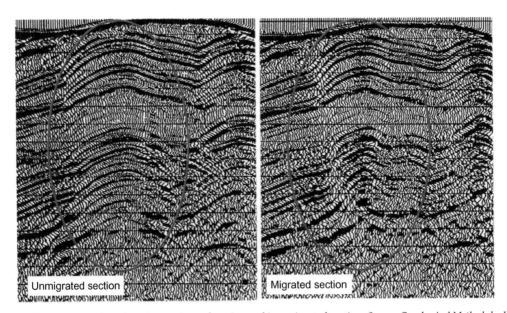

FIGURE 12.8   Anticline shape in unmigrated section and in a migrated section. *Source: Geophysical Methods by Ken Larner et al.*

Syncline

Syncline structures have a concave shape. On unmigrated seismic section, a syncline appears narrower than it really is and also all syncline will appear expanded on a migrated seismic section (Figure 12.9).

FIGURE 12.9 Syncline shape in unmigrated section and a migrated section. *Source: Geophysical Methods by Ken Larner et al.*

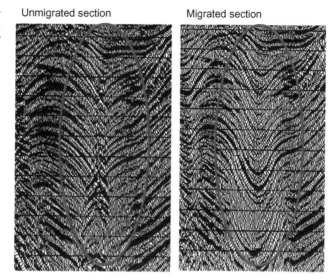

Unmigrated section   Migrated section

Incorrect velocities will lead to an incorrect shape. If the velocities are too high, then the syncline will be wider than it should be after migration. Velocities that are too low will under-migrates the structure, leaving it narrow.

Buried-Focus Anticline

If the syncline is tight there is a problem. This is because tight syncline has a seismic signature that includes the appearance of an anticline and this can be mistaken for a potential oil and/or gas trap on a seismic section. This shape that includes the appearance of an anticline is called a buried-focus anticline. In the oil and gas industry, they never drill a buried-focus anticline.

Note that the ray paths in the centre of the syncline will make a 'bow-tie' because reflections from either side of the structure crossover. This resulted to the buried-focus bow-tie or anticline and it will be corrected by migration process.

Figure 12.10 is real seismic data that show the buried-focus bow-tie effect which resulted from tight syncline.

FIGURE 12.10 Buried-focus anticline. *Source: Universidade Fernando Pessoa Porto, Portugal – Seismic Sequence Stratigraphy.*

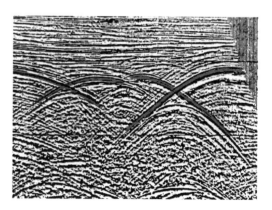

Fault

Fault affects seismic response in two ways: firstly, due to the dip of the fault plane seismic reflections from the fault plane are not located at the midpoint of the source–receiver offset but at a point up-dip on the fault plane. Thus, the seismic response for a dipping fault plane is the same as for a dipping interface. This is conceptualized in Figure 12.11.

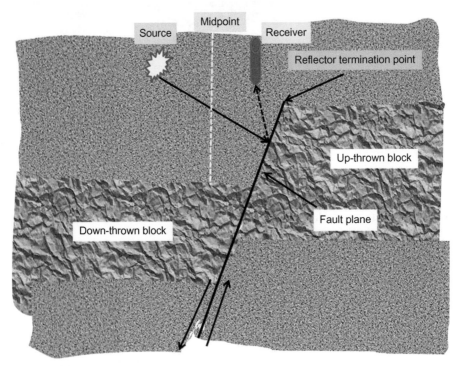

FIGURE 12.11   Fault plane reflection.

Secondly, fault can cause formation to terminate abruptly. The seismic energy directed at the formation termination point is reflected in all directions (Figure 12.12).

Since the termination point acts as a reflection surface, receivers (geophones) on the surface will record seismic waves of the reflection point just as they would of an interface. As the distance from the point increases, the travel time increases and the resulting reflection produced a diffraction curve. Diffraction curve looks like an anticline (Figure 12.13).

The apex of the diffraction curve is located at the true reflection point. This is called the diffraction point. All the reflections recorded at the diffraction curve can be attributed to this one point.

One of the objectives of the seismic migration process is to collapse all the reflected events in the diffraction curve back to the true reflection point and to restore abruptness to the faulted reflection and fault becomes more apparent on the migrated section. This is shown in Figure 12.14.

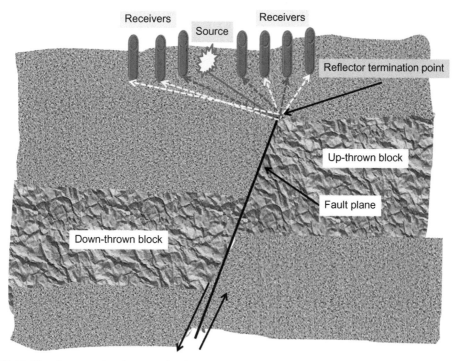

FIGURE 12.12   Conceptualized scattered reflections from reflector termination point.

FIGURE   12.13   Diffraction
curve formed as a result of the
formation termination point.

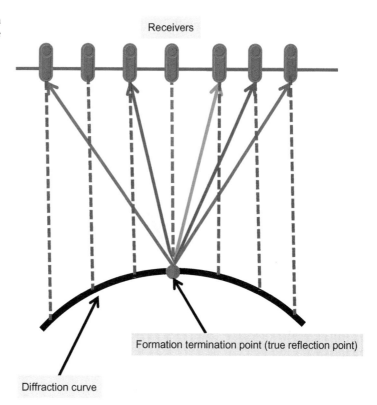

Migration also repositions an incline reflector, whether it is a stratigraphic interface or fault plane to its correct spatial location.

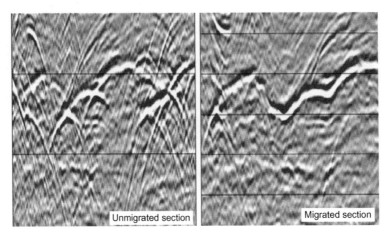

FIGURE 12.14   Unmigrated seismic section with diffraction curves and the same data after migration. *Source: Shell E&P Learning.*

In Figure 12.14 you can see the diffraction curves on the unmigrated section. On the right is the migrated section and notice that migration process has collapsed the diffractions curves and moves dipping events laterally in the up-dip direction and upwards time or earlier in time. That is, migrated image has greater dip than unmigrated image. The amount of movement depends on the dip, velocity and time of the event. If the velocity field is incorrect, the fault plane will not move to the correct place.

If the data are migrated with a faster velocity than the correct velocity, you will see diffractions on the migrated section turned into synclinal shapes called smile. If the velocities are too slow, the under-migration will leave a residual frown in the data. Migrating with the correct velocity collapses all diffractions into a point.

Note also that if there is a single point of reflected event on the unmigrated section that does not correspond to a real event, for example, a spike, this will migrate into a parabolic smile on the migrated section.

TIME MIGRATION

Time migration is the repositioning of reflected seismic data from their apparent reflection point on a midway between the source and receiver to their true reflection points in space and time (Figure 12.15).

FIGURE 12.15 Apparent reflection point (A).

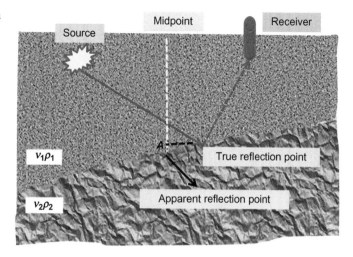

Principles of Time Migration

Time migration uses three principles: Kirchhoff migration method, finite-difference migration method and Fourier transform migration method. But we will discuss two of the time migration principles.

Kirchhoff Migration

All migration methods derive a solution to the wave equation. The wave equation is expressed below

$$\nabla^2 u = \frac{\partial^2 u}{\partial y^2} + \frac{\partial^2 u}{\partial x^2} + \frac{\partial^2 u}{\partial z^2} = \frac{1}{v^2} \frac{\partial^2 u}{\partial t^2}$$

$$\nabla^2 u = \frac{1}{V^2} \frac{\partial^2 u}{\partial t^2}$$

where u is the seismic wave field; x, y and z are the three coordinates of a three-dimensional coordinate system; t is the time; and V is the propagation velocity or rate at which seismic wave propagates.

The Kirchhoff migration uses integration to solve the wave equation. The Kirchhoff migration method uses geometry and Huggen's principle to collapse diffraction and reposition the recorded data. It considers the apex of the diffraction curve as the location of the true reflection point (Figure 12.16).

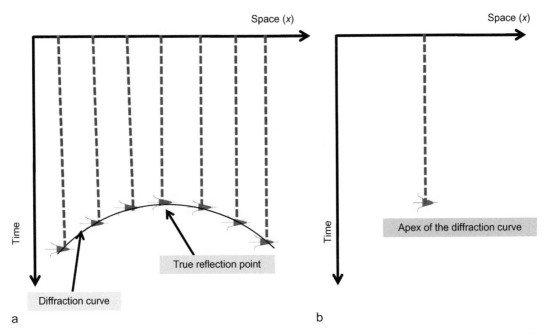

FIGURE 12.16   (a) Amplitude on the diffraction curve which resulted from the formation termination point and (b) amplitudes summation along the diffraction curve.

Kirchhoff migration collapses diffraction by summing the amplitudes along the diffraction curve and placing the sum at the true reflection point (the apex of the diffraction curve). This will put the energy belonging to the diffractor at its correct position onto the migrated trace. This process is done for every migrated trace, and the input consists of the unmigrated section.

IMPORTANT PARAMETER TO THE KIRCHHOFF MIGRATION

There are two parameters that are important to the Kirchhoff migration method.

- The first parameter is related to the diffraction curve itself. It is called the aperture width parameter. Successful Kirchhoff migration requires good diffraction definition. When the Kirchhoff migration method is used, it is important to note that the shape of the diffraction curve varies with time and distance.

   A sufficient number of traces must be included in the processing window in other to completely define the diffraction curve. If the diffraction curve is only partially recorded or if the point diffractor is so close to the surface that only a small curve is recorded, then the diffraction curve may be collapse at the incorrect point and the data may be incorrectly migrated.
- The velocity model is the second critical parameter when using a Kirchhoff migration method. Kirchhoff migration is extremely sensitive to slight changes in the velocity model. For instance, using a velocity that is 5% faster than the true velocity will result in

over-migration of the data, and also a velocity 5% slower than the true velocity will result in under-migration of the data.

Note also that Kirchhoff migration method cannot handle seismic data with low signal-to-noise ratio and data with lateral velocity variation.

The Kirchhoff migration is the most commonly used for depth processing. It is particularly attractive due to its speed and target-oriented capability, which enables efficient velocity model building and updating. In addition, the Kirchhoff migration offers adaptability to irregular acquisition geometries, flexibility in handling anisotropic and converted wave velocity models and surface topography.

Figure 12.17 shows seismic migrated image using Kirchhoff migration.

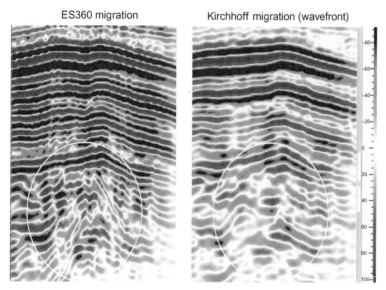

FIGURE 12.17   Migrated seismic image using Kirchhoff migration. *Source: Paradigm – Earth study 360 full-azimuth (www.paradigm.com).*

Finite-Difference Migration

The second time migration method uses differentiation to obtain a solution to the wave equation and is generally known as finite-difference methods. Finite-difference also known as downward continuation migration maps the diffraction curve (Figure 12.18) recorded at one depth to another as if it were recorded at the new depth and predicts the change in the waveform produced by the change in depth.

In other words, finite-difference migration extrapolates input data (CMP stack sections or pre-stack) using finite increments of depth and predicts what the data would look like at the new depth.

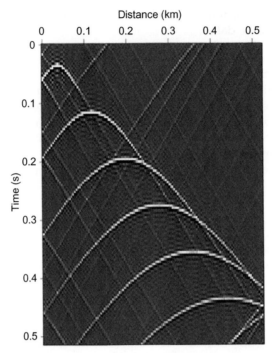

FIGURE 12.18   Downward continuation migration method. *Source: www.reproducibility.com from Jon Claerbout Book (Basic Earth Imaging).*

There are two parameters that are important to finite-difference migration method:

- The most important parameter is the size of the intervals in the downward continuation process because if the intervals are too small under-migration will occur. In this case, diffraction is incompletely collapsed and seismic event is not repositioned far enough up-dip. Also, using over small interval does not significantly improve the migration results but does require more computer time and cost.
- The velocity model is the second important parameter. Using a velocity that is too fast causes over-migration, while velocity that is too slow causes under-migration.

Note that finite-difference migration is not sensitive to velocity variations as Kirchhoff migration methods. The finite-difference migration can accommodate minor lateral velocity variations.

Finite-difference migration method can handle data with small lateral velocity variations and can also handle data with low signal-to-noise ratio. Note also that finite-difference migration takes long computing time to process.

DEPTH MIGRATION

Both the Kirchhoff migration and variations on the other wave equation techniques can be modified to take into account the bending of ray paths by refraction within a complex geology. In that case, rays are trace through the subsurface (Figure 12.19) to locate where the reflections which the geophysicists see on the seismic data actually come from. These modified techniques are usually referred to as depth migration.

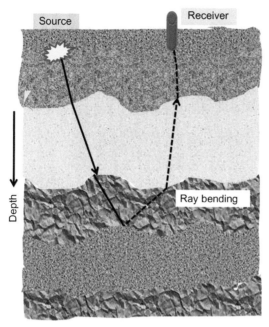

FIGURE 12.19   Conceptualized ray bending in the subsurface.

In depth migration, data are migrated in depth domain and the input data for the process is unmigrated seismic data in time domain. The output is migrated data in the depth domain. Note that conversion from time to depth requires average velocity. Average velocity can be derived from stacking velocity by first calculating the interval velocity.

Depth migration is extremely sensitive to velocity model, so detailed velocity model of the subsurface is needed. Incorrect velocity model produces inaccuracy in the migration results. However, if the velocity model is accurate, then the final results from the depth migration process are an improvement over those resulting from time migration methods.

Note that depth migration can handle both vertical and lateral velocity variations as well as moderate velocity variation. Note also that depth migration requires longer computer time to process, which in turn lead to greater data processing cost. Depth migration requires very accurate interval velocity data. Figure 12.20 shows an interval velocity model used for depth migration.

Principles of Depth Migration

There are three principles of depth migration:

- Finite-difference depth migration
- Ray-theoretical depth migration
- Image-ray tracing depth migration

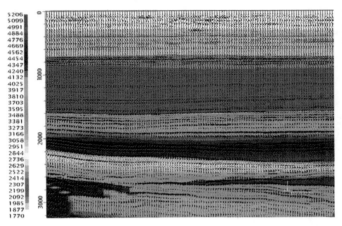

FIGURE 12.20   Velocity model used for depth migration. *Source: SpringImages (www.springerimages.com).*

DIFFERENCE BETWEEN TIME AND DEPTH MIGRATIONS

- Depth migration takes ray bending into account, while time migration does not do that.
- Depth migrations can be applied to both pre-stack and post-stack data.
- Depth migration techniques require greater computer time to run than time migration.

In the seismic sections displayed in Figure 12.21, layers corresponding to low interval velocities appear closer together in depth section than in time section. Layers corresponding to high interval velocities appear further apart in depth section than in time section.

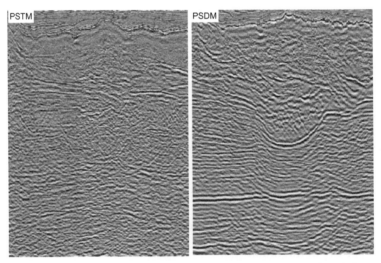

FIGURE 12.21   Pre-stack time migration (PSTM) and pre-stacked depth migration (PSDM). *Source: PGS Publication, vol. 2(4), September 2002.*

Note that conversion from time section to a depth section improves the image by producing more realistic thickness between reflectors – high velocity produces apparent small thickness in time section but depth conversion reserves this.

Note also that for the sake of consistency the final depth section is converted back into time since that's the usual scale for a seismic section.

PRE-STACK MIGRATION

Some structures are impossible to stack correctly without first migrating the data before stack. Pre-stack time and pre-stack depth migrations use all of the data, usually in common-offset planes, as the input into the migration process.

Note that gathers produced from pre-stack data are used for AVO analysis because they preserved amplitude as a function of offset.

PSTM (Pre-Stack Time Migration)

If velocity field varies only slightly laterally, pre-stack time migration (PSTM) will give a good image of geological structures in the subsurface (Figure 12.22).

FIGURE   12.22   Pre-stacked   time   migration (PSTM) section. *Source: PGS.*

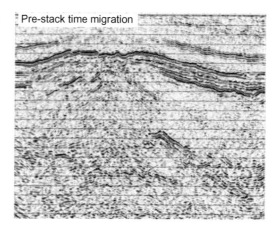

PSDM (Pre-Stack Depth Migration)

If velocity varies rapidly laterally, pre-stack depth migration is required to achieve a clear image of the geological structures in the subsurface (Figure 12.23).

POST-STACK MIGRATION

If the layers are more or less flat with small variations in velocity, a post-stack migration will provide a good image of the geological structure (Figure 12.24).

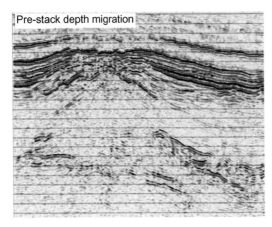

FIGURE 12.23 Pre-stacked depth migration (PSDM) of the same area as in Figure 12.22. *Source: PGS.*

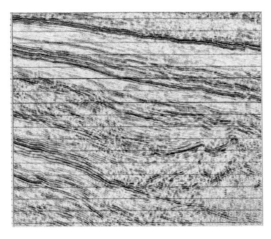

FIGURE 12.24 Post-stack migration data. *Source: www.qualityseismicservices.com.*

Note that pre-stack migration greatly improves the results if the structure within the area is highly complex. Just like depth migration, pre-stack migration requires more computer time and therefore cost more than post-stack migration.

COMPARISON BETWEEN POST-STACK AND PRE-STACK MIGRATIONS

Figure 12.25 compares post-stack migrated (left) and pre-STM 3D seismic sections. In the fractured basement with complex geology, the pre-STM-processed dataset provides much better imaging.

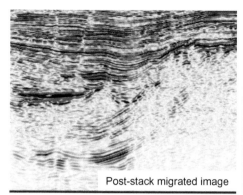
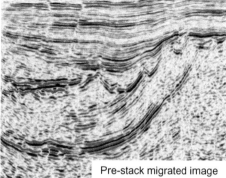

Post-stack migrated image

Pre-stack migrated image

FIGURE 12.25 Post-stack migrated section and pre-stack migrated section. *Source: GEOMEGA Petroleum Geoservices – Advanced Seismic Imaging (www.geomega.hu).*

FINAL FILTER APPLIED TO SEISMIC DATA

Filters are applied at the end of the data processing sequence to tidy-up the final stacked or migrated section. Recall that the term filtering usually means frequency filtering. Filter is used to reduce or attenuate frequency components where noise dominates over the seismic data.

Band-Pass Filter

Previous data processing steps such as de-convolution have tried to keep the bandwidth of the seismic data as wide as possible. In so doing, the de-convolution processes may have over-compensated the spectrum at early times and under-compensated it at later times and thus may have over-emphasized the noise at and beyond the signal spectrum.

Furthermore, the reflection clarity of the seismic data may change from trace to trace for a number of reasons such as different noises, reduced stack amplitude by reason of dip and trace-to-trace differences in the de-convolution. For these reasons, the final stack or migrated section needs to be filtered. Filtering out the more noise-dominated frequency components will make data look cleaner and better. The final filter applied to the data is called band-pass filter.

A band-pass filter attenuates frequencies below its low cut-off frequency and above its high cut-off frequency. The difference between these two cut-off frequencies is the pass band.

In the frequency domain, band-pass filter is the multiplication of the amplitude spectrum with the amplitude response and the addition of the phase spectrum to the phase response.

When making a bandwidth choice to apply to the data, it is important to note that low frequencies are important to understand reflection character and high frequencies are important to improve the resolution of the data. So filter with the widest bandwidth compactable with an acceptable level of noise is chosen by the geophysicists.

As an example, Figure 12.26 is used to demonstrate the effect of filter on seismic section.

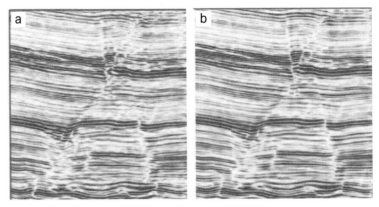

FIGURE 12.26   Seismic section before (a) and after (b) applying principal component (PC) filtering (data courtesy of Arcis Corporation). *Source: Emerging and future trends in seismic attributes by Satinder Chopra, Arcis Corporation, Calgary, Canada and Kurtd J. Marfurt, University of Oklahoma, Norman, USA, March 2008, The Leading Edge.*

Notice the cleaner background and focused amplitudes of the seismic reflections after filtering as well as the preserved fault edges.

Note that filtering has removed the noise-dominated frequency components in the unfiltered seismic data and makes the filtered section (on the right) cleaner and better.

Types of Band-Pass Filters

There are two types of band-pass filters that are commonly applied to seismic data. They are minimum-phase and zero-phase filter.

When a filter is applied to tidy-up the final stack or migrated seismic section, zero-phase filter is applied. When a filtering is applied to the seismic data at intermediate processing step, minimum-phase filter is applied, since de-convolution steps may rely on the wavelet being a minimum phase.

Minimum-phase wavelet has no amplitude at time (Figure 12.27a), while zero-phase wavelet has amplitude at time zero (Figure 12.27b).

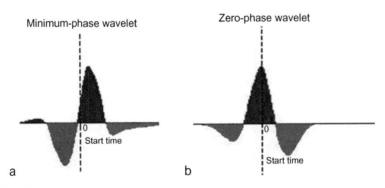

FIGURE 12.27   Minimum and zero-phase wavelet.

Zero-phase wavelet is preferred for interpretation because the strong central peak at time zero is easy to relate to the reflector of interest.

The reflection time for a zero-phase wavelet is at the central peak or trough, so picking reflection times is much easier and accurate for zero-phase wavelets. This is one reason why zero-phase wavelets are usually output in seismic data processing.

Applications of Band-Pass Filtering

There are three common applications of band-pass filtering:

- *Single band-pass filtering.* This applies a single filter to all traces at all times.
- *Time-variant filtering.* This allows the geophysicists to specify a time-varying filter which is then applied to all traces.
- *Time and spatially variant filtering.* This applies a series of time-varying filters that vary along the seismic line.

Time-Variant Filter

The frequency content of the seismic signal becomes lower as depth increases because higher frequency components of the seismic signal are absorbed as the signal propagate deeper in time. Because of these, noise in the stack or migrated section may still be broadband. For this reason, the geophysicists look for the optimum filter at various times down the section which provides a good signal-to-noise ratio. Therefore, the geophysicists need time-variant filtering. Also because the signal spectrum and the signal-to-noise ratio change with time and sometimes with space, it is more likely that one filter would not do for the entire section. So the final filter is probably time-variant.

Further Reading

Dragoset, B., 2005. A Historical Reflection on Reflections. WesternGeCo, Houston, USA.

Lorentz, M., Bradley, R., An Introduction to Migration. http://www.geol.lsu.edu/jlorenzo/ReflectSeismo197/rcbradley/WWW/rcbradley1.html.

Gadallah, M.R., Fisher, R.L., Novermber 30, 2004. Applied Seismology: A Comprehensive Guide to Seismic Theory and Application. Pennwell books. ISBN: 978-1-59370-022-5.

Al-Chalabi, M., 1974. An analysis of stacking, RMS, average, and interval velocities over a horizontally layered ground. Geophys. Prosp. 22, 458–475.

Claerbout Cecil, J.F., Green, I., 2008. Basic Earth Imaging.

Dix, C.H., 1955. Seismic velocities for surface measurements. Geophysics. 20, 68–86.

Liner, C.L., 2004. Elements of 3D Seismology, second ed. PennWell Books. ISBN: 978-1-59370-015-7.

Danbom, S.H., Ph.D., P.G., 2007. Exploration Geophysics 2, Reflection Seismic Data Processing Lecture Slide (spring).

Image (Figure 7.14) Source: Elsevier J. 160(3–4), 1 September 1999.

Image (Figure 12.26) Source: Emerging and future trends in seismic attributes by Satinder Chopra, Arcis Corporation, Calgary Canada and Kurt J. Marfurt, University of Oklahoma, Norma, USA, March 2008, The Leading Edge.

Image (Figure 12.25) Source: GEOMEGA Petroleum Geoservices – Advanced Seismic Imaging. www.geomega.hu.

Images (Figures 5.8 and 5.9) Source: Geotrace – Geometry and Refraction Statics. www.geotrace.com.

Images (Figures 5.7, 11.6 and 12.5) Source: Land Ocean Energy Services Co., Ltd. – Seismic Data Processing Software. www.idocean.com.cn.

Images (Figures 10.3, 10.5, 10.6, and 10.7) Source: Choice Geophysical. www.choicegeophysical.com.

Image (Figure 5.3) Source: Preliminary Results of a Geophysical Test of Low-Angle Dip on the Seismogenic Dixie Valley Fault, Nevada by John N. Louie, S. John Caskey and Steve G. Wesnousky. Project funded by the National Science Foundation, Tectonics Program EAR-9706255.

Image (Figure 7.9) Source: Resolution Resources International – Seismic Survey. www.rri-seismic.com.

Image (Figure 12.10) Source: Universidade Fernando Pessoa Porto, Portugal – Seismic Sequence Stratigraphy.

Image (Figure 9.17) Source: Work Flow Shown to Develop Useful Seismic Velocity Models by Khalid Amin Khan and Gulraiz Akhter. Oil and Gas Journal, 09/05/2011.

Lynn, W.S., Claerbout, J.F., 1982. Velocity estimation in laterally varying media. Geophysics. 47, 884–897.

Andeson, R.G., McMechan, G.A., Noise – Adaptive Filtering of Seismic Shot Records. University of Texas. 1987 SEG Annual Meeting, October 11–15.

Chaouch, A., Mari, J.L., 2006. 3-D land seismic surveys: definition of geophysical parameters. Oil Gas Sci. Technol. – Rev. IFP. 61 (5), 611–630.

Paradigm – Echos Seismic Processing Software. www.pdgm.com for Figure 10.8.

Prospecjiuni, S.A., Bucharest University, Str. Coralilor 20, 78449 Bucharest, Romania. July 5 – 9, 1990. Practical aspects of post-stack migration. In: Second Balkan Geophysical Congress and Exhibition. Istanbul.

Pre-Stack Migration in PGS, A Publication of PGS Geophysical, vol. 2 No. 4 September 2002.

Schultz, P.S., 1982. A method for direct estimation of interval velocities. Geophysics. 47, 1657–1671.

Schlumberger Oilfield Glossary – Common Midpoint. www.glossary.oilfield.slb.com.

Schlumberger Oilfield Glossary – Multiple Reflections. http://www.glossary.oilfield.slb.com/Display.cfm?Term=multiple%20reflection.

Sheriff, R.E., 1981. Encyclopedia Dictionary of Exploration. Society of Exploration Geophysics, Tulsa, OK.

Beasley, C., Mobley, E., 1998. Spatial de-aliasing of 3-D DMO. The Leading Edge. 17 (11), 1590–1594.

Schleicher, J., Costa, J.C., Novias, A., Time-migration velocity analysis by image-wave propagation of common-image gathers. Geophysics 73 (5), VE16–VE171, 11F1GS. SEG, Expanded Abstracts, 01/2008.

Three-Dimensional Seismic Data: Schlumberger Oilfield Glossary. www.glossary.oilfield.slb.comwww.gxtec.gemon.co.uk.

True – Azimuth 3D General Surface Multiple Prediction (GSMP), WesternGeco (Figure 10.2). www.slb.com.

Islemi, H.-Y., Yilmaz, O., 1988. Velocity-stack processing. Jeofizik. 2, 3–16.

Anderson, R.G., McMechan, G.A., Weighted stacking of seismic data using amplitude – decay rates and noise amplitudes. Geophys. Prosp. 38 (04), 365–380. Center for Lithospheric Studies, University of Texas at Dallas.

Yilmaz, O., 2001. Seismic Data Analysis. Society of Exploration Geophysicists, Tulsa. ISBN: 1-56080-094-1.

Bacon, M., Simm, R., Redshaw, T., 2003. 3-D Seismic Interpretation. Cambridge University Press. ISBN: 0-521-79203-7.

SEISMIC DATA INTERPRETATION METHODOLOGY

Understanding Seismic Interpretation Methodology

OBJECTIVE OF SEISMIC DATA INTERPRETATION

The objective of seismic data interpretation is to extract all available subsurface information from the processed seismic data. This includes structure, stratigraphy, subsurface rock properties, velocity, stress and perhaps reservoir fluid changes in time and space.

This process requires the best possible acquisition and processing to have been performed on the seismic data and also requires analogue knowledge of local geology (study area) from outcrop and pre-existing wells. A good knowledge of geologic history of the area to be studied is important in making quality decisions during interpretation of the seismic data.

It is important for the seismic interpreters to have a developed understanding of the factors influencing regional tectonic sedimentation in the basin (area) to work on.

The interpretation is normally done interactively in the industry on an interpretation workstation shown in Figure 13.1.

FIGURE 13.1   Seismic interpretation workstations. *Source: Visualization facilitates large volume, complex 3D seismic interpretation, Oil&Gas Journal. www.ogj.com.*

UNDERSTANDING THE SEISMIC DATA

Before going further lets define some important terminology that would enhance the readers understanding (Figure 13.2).

Seismic Section

A seismic section is a display of seismic data along a line, such as 2D seismic profile or profile extracted from a volume of 3D seismic data. A seismic section consists of numerous traces with location given along the x-axis and a two-way time or depth along the y-axis. Seismic section is also called seismic line (Figure 13.3).

3D seismic volume can be sliced in any vertical dimension to create 2D lines, or sliced in horizontal plane to create time slices, which represent constant time (Figure 13.4).

CDPs ⟶ seismic line

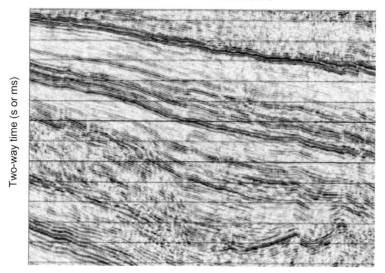

FIGURE 13.2   A post-stacked seismic section. *Source: Quality Seismic Services. www.qualityseismicservices.com.*

FIGURE 13.3   3D seismic volume. *Source: Southeast Louisiana Shallow Gas – 1: Louisiana Lagniappe: shallow gas play concept, evaluation techniques, analogs by Andy Clifford and Elizabeth Goodman. Oil&Gas Journal, 12/06/2010. www.ogj.com.*

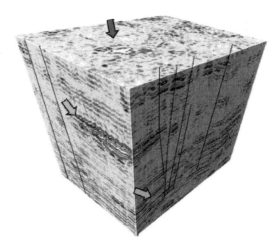

FIGURE 13.4   Time slice. (Data courtesy of Arcis Corporation.) *Source: Emerging and future trends in seismic attributes. Satinder Chopra, Arcis Corporation, Calgary, Canada. Kurt J. Marfurt, University of Oklahoma, Norman, USA, March 2008, The Leading Edge.*

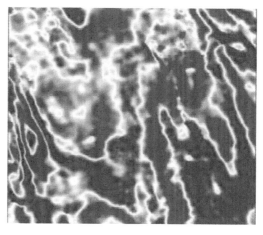

The volume can also be sliced along reflection boundaries to create 'horizon slices' (Figure 13.5).

FIGURE 13.5   Horizon slice. (Data courtesy of Arcis Seismic Solution). www.arcis.com/pages/Spectraldecom positionRS.asp.

Time Section

Seismic section is called a time section because seismic data are recorded in two-way travelled time. Figure 13.6 shows seismic section in time.

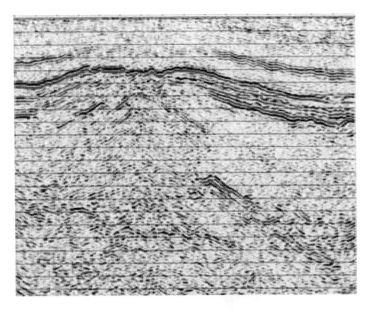

FIGURE 13.6   Shows time section. *Source: PGS*

Depth Section

Seismic section is called a depth section if the section has been converted from time to depth (Figure 13.7).

FIGURE 13.7 Shows depth section of the same area as figure 13.6. *Source: PGS.*

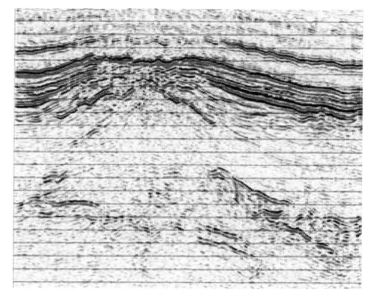

CDP

CDPs are defined as the common reflecting depth points on a reflector. CDPs are typically marked at intervals along the top of seismic lines (Figure 13.2) and they are regularly spaced to form a horizontal scale.

Post-Stack Seismic Section

Post-stack section (Figure 13.2) is made up of stacked traces measured in seconds or milliseconds.

Pick

A feature interpreted on seismic section by selecting and tracking horizon or other events. Correlation of seismic picks to geologic picks, such as formation tops interpreted from well logs, improves interpretation.

Note that an event in a seismic section can represent a geologic interface such as fault, unconformity and changes in lithology.

Two-Way Time

Two-way time (TWT) is the time required for the seismic wave to vertically travel from the source to the reflector in the subsurface and reflect back to a receiver at the surface.

Two-way time (TWT) is recorded on the vertical axis of the seismic line in seconds and it can also be express in milliseconds (Figure 13.8).

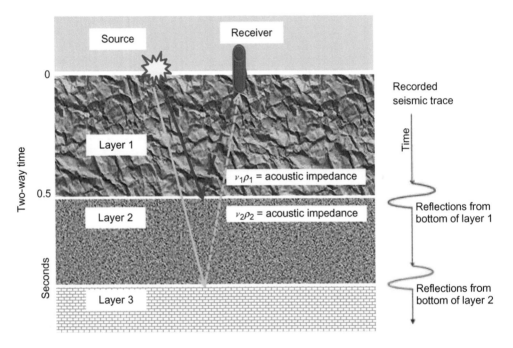

FIGURE 13.8   Conceptualized seismic reflection method.

Seismic Trace

The recorded seismic data at the surface due to the response of the earth layers to seismic source is called a trace.

Coloured Wiggle Trace

A trace whose peaks and troughs represent the acoustic softness and hardness in the earth (Figure 13.9).

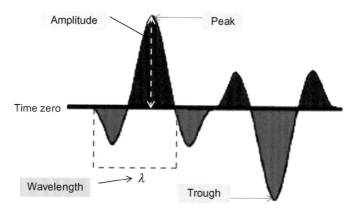

FIGURE 13.9   A trace.

Acoustic Impedance

Acoustic impedance is a layer property of a rock and it is equal to the product of compressional velocity and density.

Seismic Reflection

Seismic reflection occurs whenever there is contrast in acoustic impedance between rock layers, for example, contrast between sand and shale layers.

Seismic Amplitude

It is the maximum displacement from time zero of the seismic trace (Figure 13.9). The amplitude is a measure of how big the wavelet is, that is, the magnitude of the excursion from time zero (peak = positive) or (trough = negative).

The 'sine wave' in Figure 13.9 is a simple wavelet, that is, the shape of the acoustic wave that travels down through the earth and is reflected back up to receivers on the surface. The wavelet consists of movement that is part compression (peak or positive values as recorded by the receivers at the surface) and part rarefaction (trough or negative).

Wavelength

A wavelength (λ) is the distance over which one waveform (wavelet) is completed. It is measured in meters or feet.

$$\text{Wavelength } (\lambda) = \frac{velocity \ (v)}{frequency \ (f)}$$

Primary Seismic Reflectors

Reflectors from bedding planes and unconformities with the largest acoustic impedance.

Resolution

This is the ability to distinguish between two separate sedimentary features in a seismic section. There are two components of seismic resolution. These are vertical and lateral seismic resolution. High frequency and short wavelengths provide better vertical and lateral resolution.

To a seismic interpreter concerned with structure, the nature of the earth is described by two quantities:

- Velocity. This is because velocity defines the ray paths and the seismic reflection times.
- The acoustic impedance. This is because acoustic impedance defines the size of the reflection amplitudes.

The resolution of seismic data, that is, the thickness of stratigraphic units that can be distinguished varies by the seismic source wavelet that was used to acquire the data and the velocity of the rocks. Since velocities increases with depth, at a shallow depth of 10 m, thinner strata units can be resolved seismically. At deeper depths of about 30 m finer strata units can be resolve seismically.

Vertical Resolution of Seismic Data

Vertical seismic resolution is the minimum resolvable bed thickness. The minimum resolvable bed thickness is 1/4 of the seismic wavelength (or, two-way time thickness is 1/2 of the dominant seismic period). This is the tuning limit.

The tuning thickness is the bed thickness at which two events become indistinguishable in time, and knowing this thickness is very important to seismic interpreters who study thin reservoirs beds.

Lateral Seismic Resolution

Lateral seismic resolution means how wide a geological feature can be for it to be resolved correctly in a seismic section.

DATA SET USED FOR SEISMIC INTERPRETATION

After the geology of the area to be studied is known, a data gathering process commences.

The quality of data the seismic interpreter has determines the quality of the results of the interpretation after the study is completed.

A seismic interpretation study should access the following data sets:

- 2D or 3D seismic data cube.
- *Well data*. Well data are gotten from rock samples and measurements at depth from a wellbore. Well data required for interpretation are wire-line logs (gamma ray logs, resistivity logs, porosity log, sonic logs, density logs and neutron logs) or equivalent rock property model, check shot data, vertical seismic profile (VSP).
- Well deviation data in case the wells are deviated.

- DST (Shear sonic log) report will be needed if they are available.
- Core data from the wells in or around the study area, including any well reports, if available. Note that core data are very expensive to acquire so it may not be available at every well.

The data gathering phase requires a thorough scrutiny of the data to be used for the interpretation. This is done by carrying out a data quality check and analysis on every data set to be used. A report of the findings is made known and any grey areas are then sorted out before the project commences.

In general, this is a very incomplete listing of data necessary or used. Usually, one will integrate seismic with gravity and aeromagnetic data, take advantage of aerial photographs and outcrop maps, build a basin model and project a basin history. Seeps and geochemical analysis of the area will be carried out.

COMPARISON BETWEEN SEISMIC AND WELL DATA

- In Seismic measurement are made base on a referenced to a common datum level. Well log data are measured relative to a device on the drill rig called the Kelly Bushing (kb).
- Seismic data are recorded in two-way travel time, whereas well log data are measured in feet or metres along the wellbore.

Note that if the wellbore is not purely vertical, then we differentiated between 'measured depth' and 'true vertical depth', which has to be computed

- Seismic data sample areas and volumes within the subsurface; well log sample points along the wellbore.
- Seismic data are low frequency (5–60 Hz); a log that measures rock velocity (sonic log) uses frequencies (about 7000 Hz).
- Well log data are more direct and precise than seismic data.
- Well log data have limited resolution away from the wellbore, while seismic data have excellent resolution away from the wellbore.

WELL LOG CORRELATION

After the phase of data collection, collation, inspection and quality assurance, the wells are then correlated and appropriate sand markers are chosen to delineate the various reservoirs, shales and sands of interest.

The hydrocarbon-bearing sands (Figure 13.10) are noted, especially if the study aims to evaluate the reserve potentials of a field.

Note that correlation is the determination of the continuity and equivalence of lithologic units particularly reservoir sands or marker sealing shales across a region of the subsurface. Different geologic processes will deposit lithologic units of variable lateral continuity and developing a geologic concept of what is in the subsurface is important at this step.

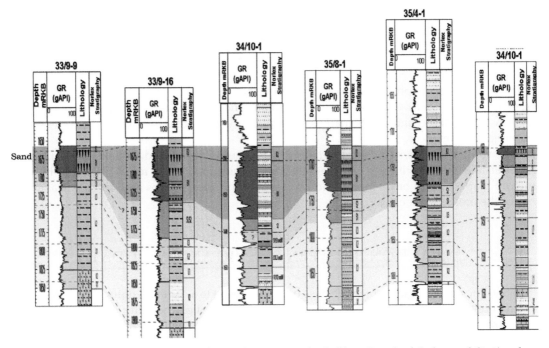

FIGURE 13.10 Shows well logs correlation. *Source: Network of offshore Records of Geology and Stratigraphy – Rogaland Group by Harald Brunstad et al.*

WELL-TO-SEISMIC TIE

After the correlation has been done, the interpreter performs a well-to-seismic tie using a synthetic seismogram. The aim is to match the measurements made through well logs, which are determined in depth domain, to the seismic data which are determined in the time domain.

What Is a Well Tie?

A well tie compares seismic data at a well location with log data from the well. To perform well-to-seismic tie requires a processed seismic data. The processed data give us a real seismic trace at the wellbore location we want to tie.

Note that we require an estimate of the seismic source wavelet from the processed data. This is needed to generate a synthetic seismic trace.

A well tie may require many logs, e.g. integrated sonic for the time–depth conversion, sonic and density for a synthetic seismic, gamma ray and resistivity log to highlight reservoir and pay zone, caliper log for borehole condition and quality control.

SYNTHETIC SEISMOGRAM

Synthetic seismogram is a seismic trace created from sonic and density logs and it is used to compare the original seismic data collected near the well location (Figure 13.11).

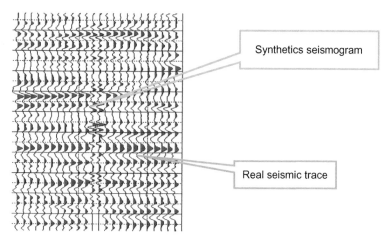

Synthetics seismogram

Real seismic trace

FIGURE 13.11   Well-to-seismic tie. The blue traces are the real seismic data. The red traces are the synthetic seismogram.

How to Generate a Synthetic Seismogram

As already stated, the primary well data required to generate a synthetic seismic trace are sonic log (inverse of the sonic log is the acoustic velocity) and density log. Check shots data from the well are also very important.

Sonic Log

Sonic logs are the principle source of well velocity data. They provide direct information about the borehole and the rocks penetrated by the drill bit. The sonic log is a measure of the time necessary for a sound wave to traverse one unit of the earth along the well bore, usually labelled 'DT' and the reciprocal of DT is the velocity (in m/s). The unit for sonic log is microseconds per foot or microseconds per metre.

Sonic logging tools measure the transit time of an acoustic wave between a down-hole source and receiver. The logging tool is an average over a couple of metres and is typically recorded two samples per foot, while the tool is pulled up along the borehole.

Density Log

The density log usually labelled 'RHOB' measures the density of the borehole and the rocks penetrated by the drill bit. The unit for density is gram per cubic centimetre.

The logs (sonic and density) measurements are usually made every 6 in. down hole, but must be aligned to be 'on-depth' with each other.

If we take aligned DT and RHOB logs and convert the depths to metres, we have a reading roughly every 15 cm.

We then multiply the sonic and density data to produce an acoustic impedance data for each reflecting interface in the subsurface.

From the acoustic impedance, we compute the reflection coefficients for each reflecting interface.

Reflection coefficient is simply the difference in acoustic impedance between stratigraphic layers divided by their sum.

That is,

$$\text{Reflection coefficient} = \frac{\rho_2 v_2 - \rho_1 v_1}{\rho_2 v_2 + \rho_1 v_2}$$

We now have a set of idealized reflection coefficients for each depth interval in the well (Figure 13.12) and these needs to be converted to time.

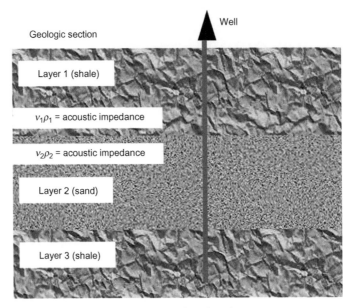

Well

Geologic section

Layer 1 (shale)

$v_1\rho_1$ = acoustic impedance

$v_2\rho_2$ = acoustic impedance

Layer 2 (sand)

Layer 3 (shale)

FIGURE 13.12   Conceptualized stratigraphy layers in the subsurface.

We assume that the sonic curve measures the P-wave interval velocities within each 15 cm (6 in.), and use this to convert the depths from the well log into time. In other words, these reflection coefficients are re-sampled from reflection depth to reflection time. This will give the relative time between reflections along a wellbore; however, few wells are logged to the surface and an absolute calibration of at least a single time to a recognized depth is often required.

How to Determine the Source Wavelet Use to Generate the Synthetic Seismic Trace

We need a source wavelet to generate a synthetic trace. Software is used to estimate the source wavelet from the processed seismic data, for a given window of the real data. Usually, this window would be at the wellbore location and near the reservoir zone of interest we aim to tie with the real seismic trace.

If there is no estimated source wavelet from the processed seismic data, we can use a standard source wavelet of a user-defined phase and frequency.

Once the source wavelet is estimated, we then convolved the reflection time series with the source wavelet to produce a synthetic seismogram (trace). The synthetic seismic trace is then compared with the real seismic trace to perform well-to-seismic tie. Note that the source wavelet represents the bandlimited character of the seismic sampling in time, since the lowest frequencies (<5 Hz – corresponding to thick geologic sequences) and the highest frequencies (>60 Hz – corresponding to very thin geologic sequences) are not represented.

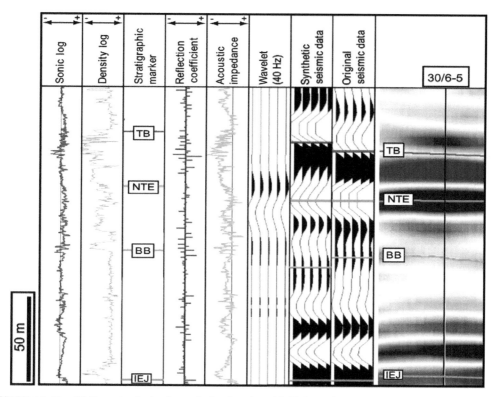

FIGURE 13.13   Well-to-seismic tie. *Source: J. Geophys. Soc., 167(6), December 2010, 1225–1286.*

Note that since the seismic source wavelet changes in the earth with depth due to absorption and attenuation, several synthetic traces may be generated based on different estimations of the source wavelet – one for shallow targets and another for deeper targets.

Well-to-seismic tie enable us to correlate the stratigraphy sequence drilled in the well to the acoustic reflection recorded on the seismic section (Figure 13.13). This enables us to tie borehole formation tops identified in the well with specific reflectors on the real seismic section and to quantitatively evaluate seismic attributes, such as amplitude.

If we obtain a good match between the synthetic seismic trace and the real seismic trace, we would be able to extract various seismic attributes to predict rock and fluid properties.

UNDERSTANDING VELOCITY

Seismic data are acquired in two-way time, whereas well data are acquired in depth. In order to tie seismic events which are in time to well data, seismic data must be converted to depth.

Time–depth correlation is based on a table that ties seismic events in time with formation tops in depth.

The table is usually presented as an interval velocity profile derived from travel time differences between adjacent time–depth pairs.

Before going further to explain time-to-depth conversion technique, let's discuss average and interval velocity.

Average Seismic Velocity

Average seismic velocity is twice the depth or distance the seismic wave travelled from the source to the reflector of interest within the earth divided by the two-way recorded travelled time

$$V_a = \frac{2Z_1}{T_1}$$

where V_a is average velocity, Z_1 is depth to the first reflector and T_1 is the two-way time.

The two-way time is the time required for the seismic wave to travel from the source to the reflector of interest and back up to the receiver at the surface, and is what is recorded on all reflection seismic data.

Two-way time (TWT) is recorded on the vertical axis of the seismic line in seconds. Sometimes it is more convenient to express time as milliseconds.

Note that $1\ s = 1000\ ms$.

Two-way time (TWT) does not always equate directly to depth because wave paths are not always vertical. Depth of a specific reflector can be determined using boreholes. For example, $4500\ m = 1.15\ s$. TWT would be a time–depth pair.

How to Determine an Average Seismic Velocity Curve

Let Figure 13.14 represents the interpreted seismic section of our study area with well at shot point 500. From a synthetic seismogram, we correlate a well at 4500 ft to the shallow horizon, with a seismic reflection time of 1.15 s at this location. What is the average velocity to the shallow horizon?

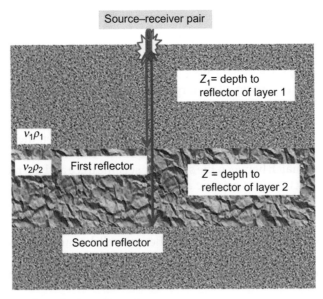

FIGURE 13.14   A zero offset seismic configuration.

SOLUTION

The total distance travelled by the seismic energy is twice the well depth (2×4500) divided by the two-way time (1.15) to the shallow horizon. This gives us an average velocity of 7826 ft./s. Note that this then represents the average velocity over the first 4500 ft of the earth above the well (Figure 13.15).

FIGURE 13.15   This set-up is used to described average velocity. *Source:* *www.publications.iodp.org.*

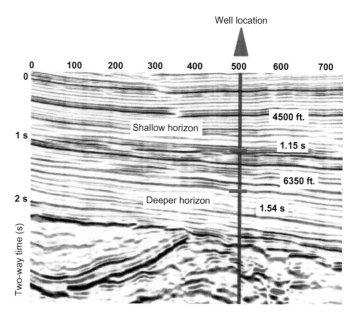

Well location

That is,

$$V_a = \frac{2Z_1}{T_1}$$

$$V_a = \frac{2 \times 4500}{1.15} = 7826\,\text{ft./s}$$

On the seismic section in Figure 13.15, at shot point 500, the deeper horizon has a reflection time of 1.54 s, which the interpreter correlates to a well depth of 6350 ft. What is the average velocity to the deeper horizon?

SOLUTION

The total distance travelled by the seismic energy is twice the well depth (2×6350) divided by the two-way time (1.54) to the deeper horizon. This gives us an average velocity of 8247 ft./s

$$V_a = \frac{2Z_2}{T_2}$$

$$V_a = \frac{2 \times 6350}{1.54} = 8247 \, \text{ft./s}$$

For the shallow horizon, the calculated average velocity of 7826 ft./s corresponds to a depth of 4500 m.

For the deeper horizon, the calculated average velocity of 8247 ft./s corresponds to a depth of 6350 m.

Let us assume that the velocity of the rocks at surface is 5550 ft./s, then we plot the two calculated average velocity against their corresponding depth. We then draw a curve that fit the data (Figure 13.16).

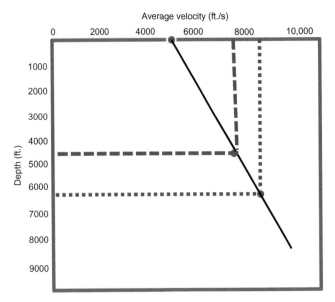

FIGURE 13.16   Average velocity curve.

Note that the average velocity to the shallow horizon is 7826 ft./s and the average velocity to the deeper horizon is 8247 ft./s. Looking at both velocity values, notice that average velocity increases with increase in depth. This is because rock velocities generally increase with depth due to higher pressure and more compaction.

Seismic Interval Velocity

Interval velocity is defined as the thickness of a stratigrahpic layer divided by the time it takes to travel from the top of the layer to its base.

The interval velocity is also equal to twice the interval thickness divided by the two-way travelled time:

$$V_i = \frac{2\Delta z}{2\Delta t} = \frac{2\Delta z}{\Delta T}$$

FIGURE 13.17   Interval velocity.

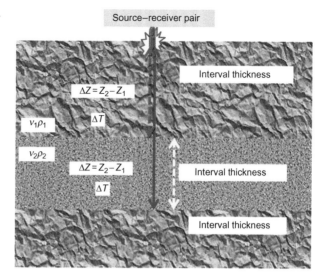

Note that the layer thickness is equal to the isopach value of the interval (Figure 13.17).

Small letter t means one-way travelled time, while big letter T means two-way travelled time.

How to Determine Seismic Interval Velocity Curve

In Figure 13.18, the depth from the surface to the shallow horizon is 4500 ft. This is also the interval thickness of the shallow layer. The two-way travelled time to the shallow horizon is 1.15 s, which is also the interval time.

FIGURE 13.18   This set-up is used to described how to determine interval velocity. *Source: www.publica tions.iodp.org.*

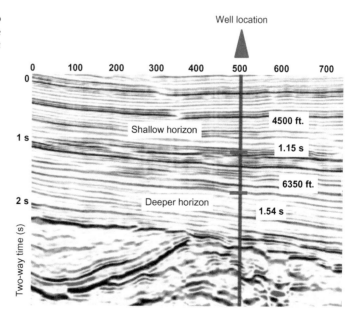

Therefore, the interval velocity of the shallow horizon is 7826 ft./s, that is,

$$V_i = \frac{2\Delta z}{\Delta T}$$

$$V_i = \frac{2(Z_2 - Z_1)}{T_2 - T_1}$$

$$V_i = \frac{2(4500 - 0)}{1.15 - 0} = 7826\, \text{ft./s}$$

For the deeper horizon, the interval velocity is twice the interval thickness divided by the difference in two-way time.

That is,

$$V_i = \frac{2(6350 - 4500)}{1.54 - 1.15} = 9487\, \text{ft./s}$$

For the shallow horizon, the calculated interval velocity is 7826 ft/s, which corresponds to a depth of 4500 ft.

For the deeper horizon, the calculated interval velocity is 9487 ft/s, which corresponds to a depth of 6350 ft.

Let's plot the calculated interval velocity values on the velocity versus depth graph and compare the average velocity curve to the interval velocity curve (Figure 13.19).

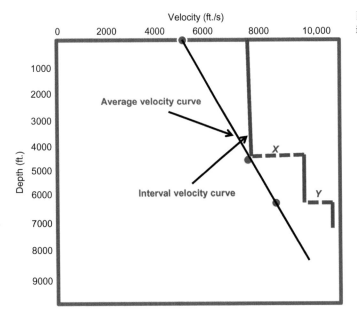

FIGURE 13.19   Average velocity and interval velocity curves.

The discrete boundaries in the interval velocity curve (X and Y) indicate stratigraphic and velocities differences between adjacent layers. Note that both the average and interval velocities curves were derived from the same data but we are measuring velocities in different ways. In practice, which of these curves we use depend on how we wish to use the velocity of the study area.

TIME-TO-DEPTH CONVERSION TECHNIQUE

There are a variety of techniques used for time-to-depth conversion of seismic data. They are constant function technique, timeline technique and average velocity technique. We will use the timeline technique to illustrate how time-to-depth conversion is applied in the oil and gas industry.

The timeline technique is used to convert time to depth for a single, interpreted horizon that reflects a stratigraphic unit.

This technique requires more than one time–depth data point for each horizon of interest so that the interpreter can statistically define a time–depth function.

To calculate an average velocity curve for the horizon of interest, the seismic interpreter uses input data from a variety of sources such as seismic-to-well tie, check shot survey, vertical seismic profiles (VSP) and integrated sonic logs.

Let's demonstrate how the timeline technique is used to convert time to depth.

In Figure 13.20, at shot point 300, the seismic reflection time of 1.54 s correlates to a well at depth of 6350 ft.

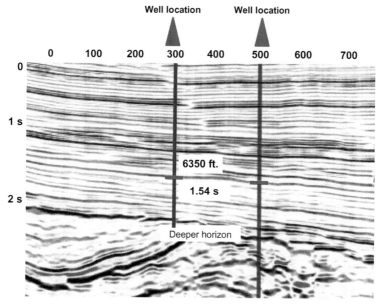

FIGURE 13.20   Timeline technique. *Source: www.publications.iodp.org.*

Let's assume we have a second well at shot point 500, and the seismic reflection time of 1.565 s correlates to a well at a depth of 6421 ft. at this point (Figure 13.21). What would the depth to the top of a potential reservoir be at shot point 700?

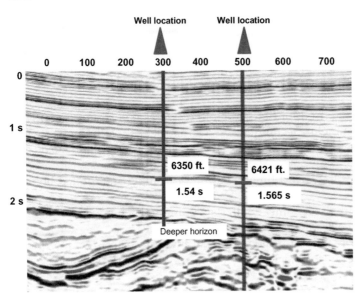

FIGURE 13.21 Timeline technique. *Source: www.publications.iodp. org.*

Solution

To calculate the depth, firstly, the seismic interpreter plots the two data points: 6350 ft. at 1.54 s and 6421 ft. at 1.565 s on the time versus depth curve. This is shown in Figure 13.22.

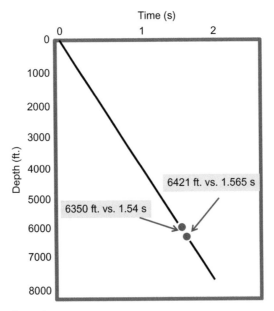

FIGURE 13.22   Time–depth graph.

The interpreter then defines the timeline by fitting a curve (in this case a straight line) to the time–depth data.

On the seismic section in Figure 13.23, the interpreter measures the depth to the interpreted horizon at shot point 700 as 1.615 s.

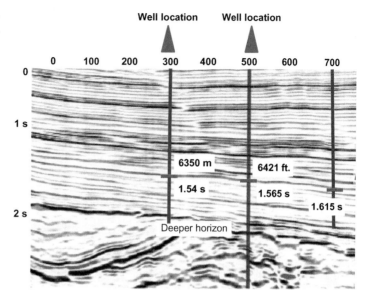

FIGURE 13.23 Timeline technique. *Source: www.publications.iodp.org.*

The seismic interpreter, then, plots the time of 1.615 s and determines the corresponding depth to the top of potential reservoir to be at a depth of 7000 ft. (Figure 13.24).

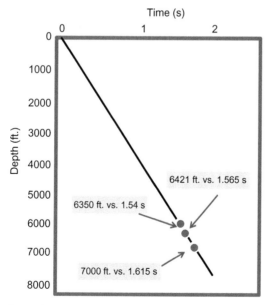

FIGURE 13.24 Time–depth graph.

FAULT PICKING

Faults are picked on the seismic section (Figure 13.25) to delineate the geological structural trend in the study area. This is carried out with the nature of the geology of the basin, that is, structural style of the area in mind (Figure 13.26).

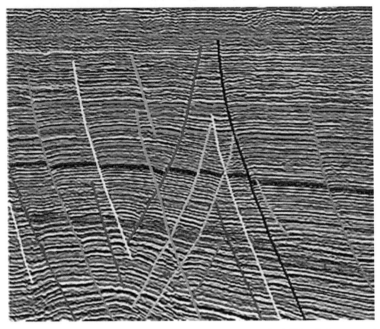

FIGURE 13.25　Interpreted fault in the seismic section. *Source: How to find Oil and Gas. www.sjvgeology.org.*

Structural styles often provide a broad context for understanding the pattern of faulting that may be expected in a region. Its basic utility lies in identifying certain basic patterns of deformation that are repeated in geologic provinces.

Examples of structural styles found in the Niger Delta basin, Nigeria, include growth fault, major bounding fault, rollover anticline, collapse crest structures, synthetic and antithetic fault, etc. To learn more about these different types of faults (Figure 13.26), go to Chapter 1 of this practical handbook.

Fault Interpretation

What are some of the clues to look for on seismic data to recognize faults? They are listed here:

- Termination of reflections
- Offset in stratigraphy markers
- Abrupt changes in dip, abrupt changes in seismic patterns - e.g. a strong, continuous reflection turns into a low amplitude region,

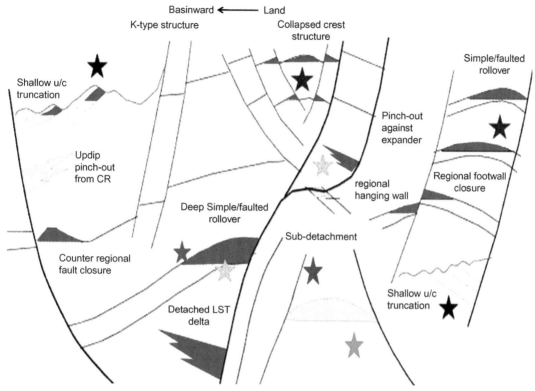

FIGURE 13.26   Structural style of the Niger Delta, Basin, Nigeria. *Source: Shell E&P.*

- Fault plane reflections – only when fault dips less than about 30
- Associated folding
- Coherency

Faults are recognized most often on a seismic section by the terminations of the strata reflections at the fault (Figure 13.27). These reflections terminations are not abrupt.

When the fault throw is small enough that the displaced horizon can be identified on both sides of the fault, the fault throw can be measured by noting the difference of travel time between the two horizons.

Note that the best measure of the throw of the fault is given by the two strata reflections. And the best indication of the fault location comes from the fault plane reflections.

When reflectors terminate abruptly the corresponding reflections tailed away smoothly into diffraction curves. As a result, the actual fault location is often difficult to pinpoint on an unmigrated seismic section (Figure 13.28).

Note that reflection terminations can be used to identify and mark unconformities. Changes in the characteristics of a reflection such as continuity, frequency, and amplitude, indicate changes in depositional facies.

Seismic migration technique collapses the diffraction curves and the faults become more apparent on the migrated seismic section (left of Figure 13.28).

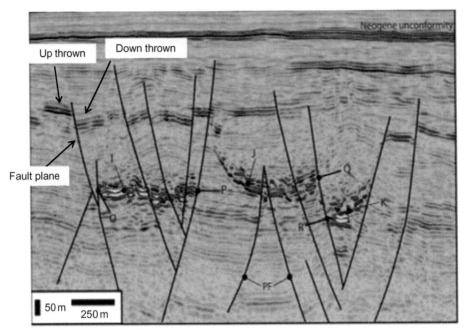

FIGURE 13.27   Seismic line showing faults. *Source: J. Struct. Geol., 46, January 2013, 186–196.*

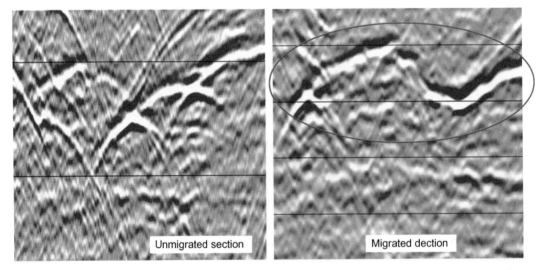

FIGURE 13.28   Unmigrated seismic section with diffraction curves and the same data after migration. *Source: Shell E&P.*

It may be difficult to identify a fault plane reflection on a seismic section due to ray-path problems around the fault and also because many faults have no clear breaks and no diffractions (Figure 13.29). These faults are recognized by looking for an abrupt change in dip of several reflectors that changes slants across the section, this could be a fault plane.

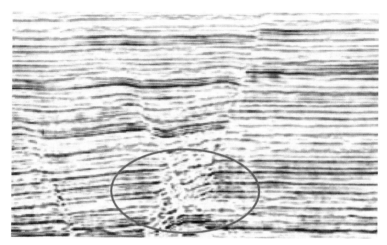

FIGURE 13.29   Fault in the seismic line. *Source: Visualizing 3D seismic data by Huw James, First break volume 27, March 2007.*

Note that seismic migration techniques also laterally reposition all inclined reflectors, whether they are stratigraphic interfaces or fault planes, to their correct spatial location.

Guild to Fault Picking

- The seismic interpreter must have a good knowledge of the regional tectonic framework of the study area and apply the structural styles to the basin.
- Seismic interpreter must use the unmigrated and migrated section when interpreting faults.

The migrated section is used to clarify and locate the fault where the section is perpendicular to the fault.

On 2D seismic data, the unmigrated section is used to check the ties on the intersections:

- The seismic interpreters must focus its attention on the seismic lines perpendicular to the major faults. Typically the dip lines.
- The direction of the fault is obtained by correlating the fault from line to line. The interpreters ensure that the faults correlated are of the same type (coherency), stress mechanism and compactable throw.
- For 2D seismic data, the interpreter must tie faults picks at the line intersection on the unmigrated section.
- The interpreters anchor the interpretation at the places where there is no dip (Figure 13.30).

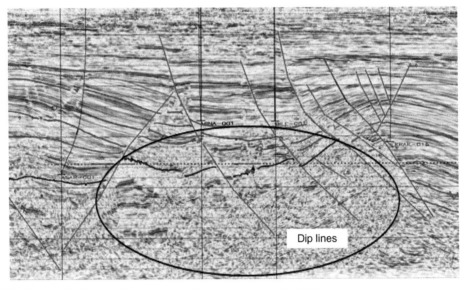

FIGURE 13.30   Dip line correlation on the seismic line. *Source: Shell E&P.*

FAULT SLICE

A fault slice is a projection of data along the surface parallel to an interpreted fault plane and displayed as a vertical section (Figure 13.31 b).

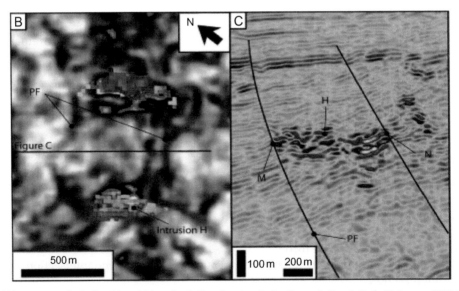

FIGURE 13.31   A fault slice (b) and (c) a seismic line showing faults. *Source: J. Struct. Geol., 46, January 2013, 186–196.*

Uses of Fault Slice

- It is helpful in mapping detailed structure along the fault plane.
- It can be used to determine the sealing and leaking properties of a fault.
- It can be used to determine the throw at any point along the fault plane.
- It can be used to plan deviated wells.

SEISMIC HORIZON INTERPRETATION AND MAPPING

Seismic horizon interpretation implies picking and tracking laterally consistent seismic reflectors with the aim to detecting hydrocarbon accumulations, delineating their extent and calculating their volumes.

There are four basic steps in seismic horizon interpretation:

Picking: To identify and follow the horizon (seismic reflector) to be mapped.
Timing: To measure the reflection time from datum to the picked reflection (horizon).
Posting: To transfer the measured reflection times (horizon) to the map.
Contouring: To show the structure, relief and closure of the chosen horizon.

WHAT IS A HORIZON?

To the exploration geophysicists, a horizon is an event, a reflection, in the seismic data; something you could pick with an automatic tracking tool. The quality is subject to the data itself. A change in the data, or seismic processing, may change the horizon.

This definition of horizon is devoid of geology. It does not match geological tops at the wells and it is not supposed to.

To the geologist, a surface picked on well logs or core data, and then interpolated between data points, can be thought of as a horizon.

SEISMIC HORIZON MAPPING

In horizon mapping, the interpreter aims to map laterally consistent geologic structures, stratigraphy and reservoir architectures.

The best place to start horizon mapping is at the top of the section, where definition is usually best. And then work down the section towards the zone where the signal-to-noise ratio is reduced and the reflector definition is less clear.

Figures 13.32–13.37 illustrate, in practice, how a horizon is mapped and interpreted.

The well in Figure 13.32, which is located near the intersection of lines 69 and 70, is used to tie seismic reflectors to known geological horizons identified in the well:

- Base Permian at 150 ms
- Blackshale Coal at 240 ms
- Near Top Dinantian at 500 ms

The potential reservoirs are Namurian and Westphalian (Upper Carboniferous) sandstones that occur below the Blackshale Coal and above the Near Top Dinantian (Lower Carboniferous) horizon.

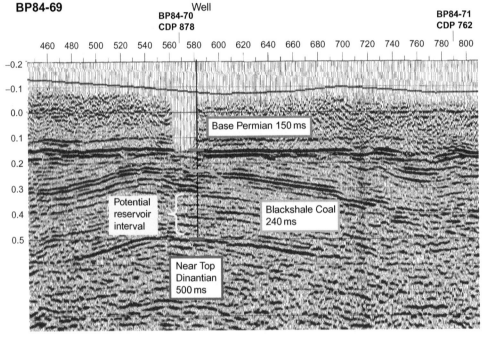

FIGURE 13.32   This set-up (well and seismic) is used to illustrate horizon mapping. *Source: UK Onshore Geophysical Library, Martin Whiteley and Dorothy Satterfield.*

The interpreter starts at the top of the section and interprets the Base Permian unconformity (the first horizon picked) away from the well on line 69 (Figure 13.33). Note that reflection terminations can be used to identify and mark unconformities.

Then, the interpreter fold line 70 at the intersection with line 69 and then match them up. This is shown in Figure 13.34.

Next, the interpreter finds and interprets the Base Permian unconformity away from well 69 and correlates it with intersecting lines 72 and 73, ensuring that the interpretation is consistent and geologically reasonable (Figure 13.35).

Finally, the interpreter unfolds line 70 (Figure 13.36).

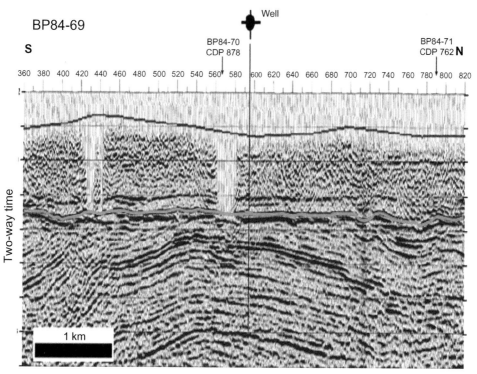

FIGURE 13.33   The mapped Base Pernain unconformity. *Source: UK Onshore Geophysical Library, Martin Whiteley and Dorothy Satterfield.*

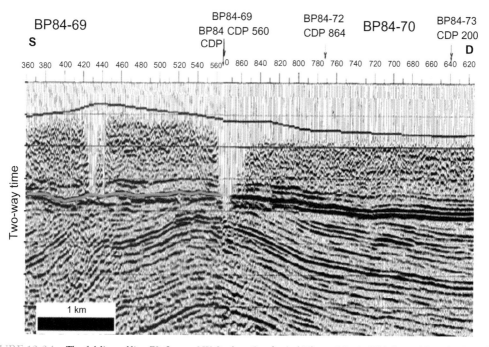

FIGURE 13.34   The folding of line 70. *Source: UK Onshore Geophysical Library, Martin Whiteley and Dorothy Satterfield.*

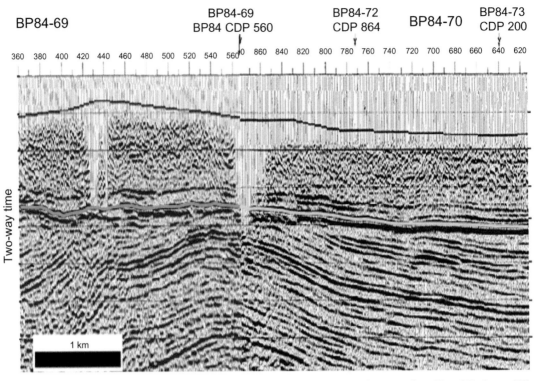

FIGURE 13.35   The folding of line 70 and the interpreted Base Permian unconformity to lines 72 and 73. *Source: UK Onshore Geophysical Library, Martin Whiteley and Dorothy Satterfield.*

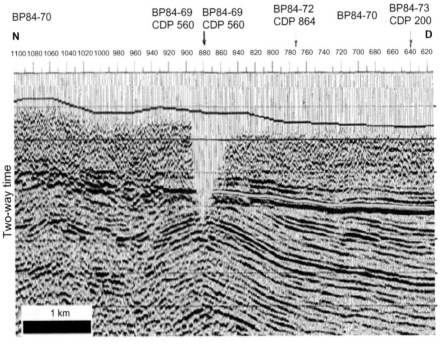

FIGURE 13.36   Line 70 unfolded. *Source: UK Onshore Geophysical Library, Martin Whiteley and Dorothy Satterfield.*

And the interpretation is completed (Figure 13.37).

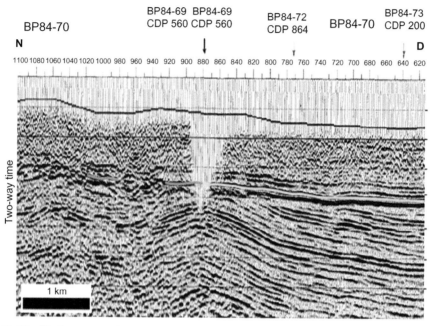

FIGURE 13.37   Horizon mapped and interpretation completed for the Base Permian unconformity. *Source: UK Onshore Geophysical Library, Martin Whiteley and Dorothy Satterfield.*

Note that this example is strictly a 2D seismic data, as one map onto a cross-line. In 3D, we often use automated technique if the cross-line dip is not too large, and if the section is not too faulted.

HOW TO GENERATE TIME STRUCTURE MAP

To generate a time structure map, the interpreter determines the reflection time values (in milliseconds) from datum to the picked Base Permian horizon at an appropriate CDP interval (Figure 13.38) and plots these values on the map (Figure 13.39). For example, on line 69 the interpreter could start by plotting values at CDP 500, 600, 700, 800 and so on.

Note that the map shows the location of the seismic lines (section).

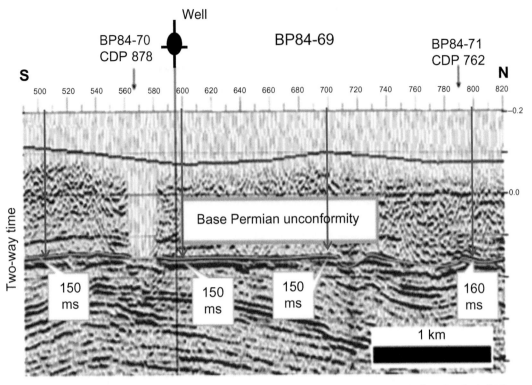

FIGURE 13.38 The picked reflection time for the Base Permian horizon. *Source: UK Onshore Geophysical Library, Martin Whiteley and Dorothy Satterfield.*

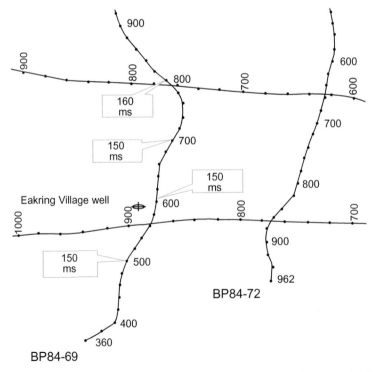

FIGURE 13.39 Seismic survey location map. *Source: UK Onshore Geophysical Library, Martin Whiteley and Dorothy Satterfield.*

The above process (the technique used to map the Base permain horizon from figure 13.32 to 13.36) is repeated for both Blackshale Coal and Near Top Dinantian reflectors (Figure 13.40).

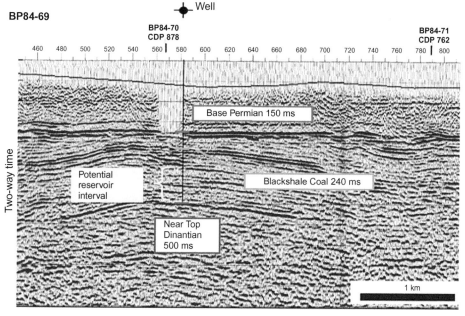

FIGURE 13.40 Blacshale Coal and Near Top Dinantain reflectors. *Source: UK onshore Geophysical Library, Martin-Whiteley and Dorothy Satterfield.*

The reflection time (in milliseconds) values for both Blackshale Coal and Near Top Dinantian horizons at an appropriate CDP interval are determined. For example, on line 69 the interpreter could start by plotting values at CDP 500, 600, 700, 800 and so on.

These time values in milliseconds are then plotted on the map against the appropriate CDPs to make a time structure map (Figure 13.41).

FIGURE 13.41 Time structure map.

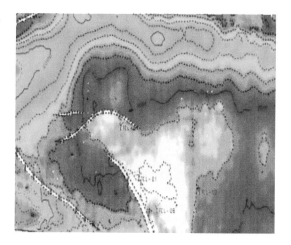

The whole field of interest is mapped on the seismic section. Depending on what the interpreters hope to achieve with the interpretation, amplitude extraction maps can be generated (Figure 13.42).

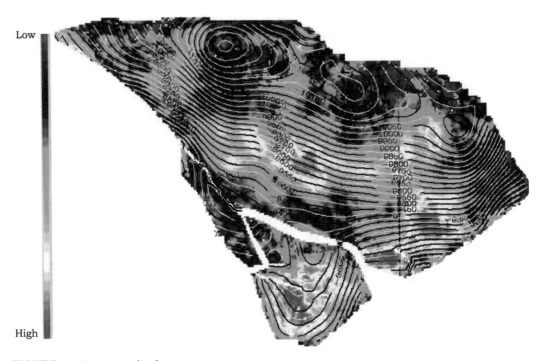

FIGURE 13.42   An amplitude contour map.

The purpose of time-to-depth conversion is to transform the subsurface time map derived from seismic horizon interpretation to an accurate depth map in which the vertical and horizontal positions as well as the size of the subsurface structure are not altered.

DEPTH CONVERSION

Seismic data are acquired in two-way travel time. Depth conversion is used to convert the two-way seismic travel time to actual depth, to provide a picture of the subsurface.

Note that interval velocity models are used for depth conversion because they can be compared to stratigraphy. Also, the two-way time value exactly where each geological layer crosses the well path is converted to one-way time.

Depth conversion integrates several sources of information about the subsurface velocity to derive a three-dimensional velocity model:

- 'Well tops', i.e. depth of geological layers encountered in oil and gas wells.
- Velocity measurements (acoustic velocity, check shot, VSP data).

- Empirical knowledge about the velocities of the rocks in the study area.
- Root mean square (RMS) stacking velocities which are derived from the processing of the seismic reflection data.

The conversion permits the production of depth and thickness maps that depict subsurface layers that are based on reflection data.

On the depth structural map, drillable prospects are identified on the map and colours are coded as it relates to oil (red colour), gas (green colour), water (blue colour) or condensate as the case may be (Figure 13.43).

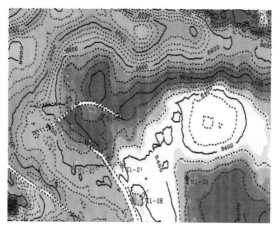

FIGURE 13.43   A depth structure map.

These (time and depth) maps are crucial in hydrocarbon exploration because they permit the volumetric evaluation of oil and gas in place, and thereby hydrocarbon reserve estimations can then be carried out on the study area.

Understanding Reflection Coefficient

Seismic reflection occurs whenever there is contrast in acoustic impedance (product of velocity and density) between rock layers (Figure 14.1). The amount of energy that is reflected is a function of the magnitude of the impedance change across a boundary. A small change in impedance results in a small amount of reflected energy; a large change in impedance results in a larger amount of reflected energy.

We can calculate a parameter called the Reflection Coefficient (RC) using a formula

$$Reflection\ coefficient = \frac{\rho_2 v_2 - \rho_1 v_1}{\rho_2 v_2 + \rho_1 v_2}$$

The reflection coefficient is the reflection response of the layered earth. It is also called the earth reflectivity series. An increase in impedance results in a positive reflection coefficient. A decrease in impedance results in a negative reflection coefficient.

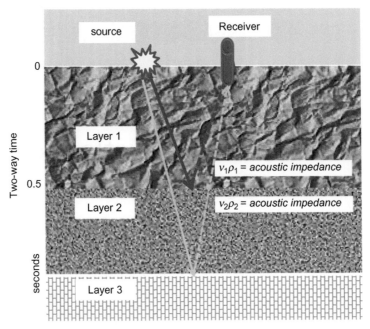

FIGURE 14.1   Described reflection coefficient.

REFLECTION COEFFICIENT AT NORMAL INCIDENCE

At normal incidence (source and receiver are in the same position, as in Figure 14.2), when a seismic P-wave (compressional wave) crosses the boundary between two layers of differing velocity and density, no reflected or refracted S-wave (shear wave) is generated. There, all angles are zero, and the non-reflected rays are merely the transmitted rays.

At normal incidence, the reflection coefficient is defined in both sign and magnitude by the impedances above and below the interface.

To a seismic interpreter concerned with structure, the nature of the earth is described by two quantities:

- *Velocity*. This is because velocity defines the ray paths and the seismic reflection times.
- *The acoustic impedance*. This is because acoustic impedance defines the size of the reflection amplitudes.

Note that the velocity is a principal stratigraphic factor in both the acoustic impedance and the reflection coefficient. Rock properties that affect velocity are porosity, pore fluid, compaction, stress, lithology and geological age.

Rock properties are the physical characteristics of reservoir rocks that enable them to store fluids and allow fluids to flow through them. The main properties of interest are rock porosities and permeabilities.

Source–receiver pair at normal incidence

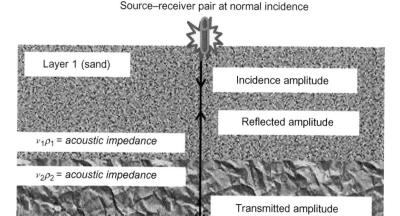

FIGURE 14.2   Reflection coefficient at normal incidence.

REFLECTION COEFFICIENT AT AN ANGLE

When a P-wave is incident on a boundary between two layers at an angle, there will be a reflected P-wave and a refracted P-wave. There are also two new waves generated: a reflected S-wave and a refracted S-wave (Figure 14.3).

The P- and S-wave velocities in layer 1 are V_{P1} and V_{S1}, respectively. P- and S-wave velocities in layer 2 are V_{P2} and V_{S2}, respectively.

At non-normal incidence, where the reflection coefficient varies with the angle, the reflection coefficient depends on

- The angle of incidence.
- The velocities below and above the interface as well as their impedances because of refraction.
- The 'sideways' or shear properties of the rocks, that is, the relative readiness of the rocks to change shape as well as volume.

Note that the reflected amplitude depends on the contrast in Poisson's ratio across the interface as well as the acoustic impedance changes.

Aki and Richards (1980) and Shuey (1985) give an approximation relation of how the reflection amplitudes vary with the rock properties. The approximation at a small reflection angle is given by

$$R(\theta) = R_0 + G \sin^2 \theta$$

where $R(\theta)$ is the reflection coefficient at the incidence angle θ, R_0 is the reflection coefficient at zero-offset (normal incidence) and G is the gradient of the amplitudes against sine of the angle

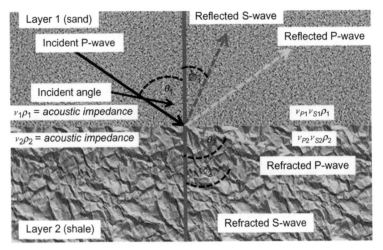

FIGURE 14.3   Reflection coefficient at an angle and mode conversion.

of incidence squared. G describes reflection behaviour at intermediate offsets and (θ) is the angle of incidence.

Note that the above equation can be used to model the seismic response when the rock properties are known.

POISSON'S RATIO

Poisson's ratio is the readiness of a compressed material to bulge. Fluids, which bulge immediately, offer no resistance to change of shape, and therefore have high values of Poisson's ratio. The value of Poisson's ratio varies from 0 to 0.5. Poisson's ratio for fluids is ~0.5, while solids have low values of Poisson's ratio.

Poisson's ratio is given by

$$\sigma = \frac{0.5 - (V_P/V_S)^2}{1 - (V_S/V_P)^2}$$

In AVO and reservoir characterization studies, V_P/V_S is sometimes used instead of Poisson's ratio.

Thus, to determine the reflection coefficient at non-normal incidence, geoscientists need to know not only the two compressional velocities (V_{P1} and V_{P2}) and the two densities (ρ_1 and ρ_2) but also either the two shear velocities (V_{S1} and V_{S2}) or Poisson's ratio. Remember that the shear properties of a rock are important to understand AVO (amplitude variation with offset). How much the amplitude of the reflected compressional wave changes with the angle is directly influenced by how much energy is converted to shear waves at the interface (Figure 14.3).

Note that AVO is use to determine the seismic response due to hydrocarbon effects across the field. Figure 14.4 shows amplitude variation at various incident angles.

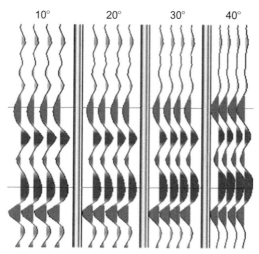

FIGURE 14.4   Shows AVO effect in the seismic traces at various angle of incidence. *Source: Shell International.*

Note also that at normal incidence, the magnitude of the reflection coefficient from soft rock to hard rock is the same as from hard rock to soft rock, whereas at non-normal incidence, the magnitude of the reflection coefficient is generally different from soft rock to hard rock and from hard rock to soft rock.

In Figures 14.5 and 14.6, as the seismic wave travels from layer 1 to layer 2, its amplitude is proportional to the magnitude of the reflection coefficient and its polarity corresponds to the sign of the reflection coefficient.

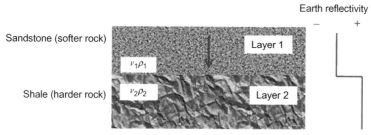

FIGURE 14.5   Positive reflection coefficient.

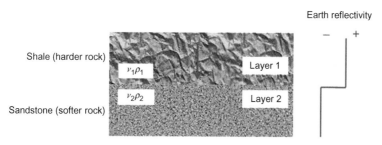

FIGURE 14.6   Negative reflection coefficient.

III. SEISMIC DATA INTERPRETATION METHODOLOGY

Both velocity and density increase with depth. Therefore, if the density is proportional to velocity, then for interfaces where velocity increases across the interface, the reflection coefficient will be positive (Figure 14.5). That is, if layer 2 is acoustically harder (high acoustic impedance) than layer 1 (low acoustic impedance), the reflection coefficient is positive. But if layer 1 is acoustically harder rock than layer 2, the reflection coefficient is negative.

UNDERSTANDING SEISMIC AMPLITUDE

In seismic acquisition, what the seismologists actually measure are reflection amplitudes. One of the key properties of each stratigraphy layer in the earth is their acoustic impedance, which is the product of velocity and density. The reflection amplitudes depend on the differences between acoustic impedances from layer to layer in the subsurface.

The strength of the reflected amplitude depends on the difference between the velocities in the layer above and below, with a stronger reflector occurring when the difference is greater. The greater the strength, the larger the swing of the trace will be; in other words, it has a bigger amplitude (Figure 14.7).

Note: A reflector is a boundary between beds with different properties. There may be a change of lithology or fluid fill from layer 1 to layer 2. These property changes cause some sound waves to be reflected towards the surface (Figure 14.7).

The amplitudes of seismic traces are often used to make a variety of geologic interpretations.

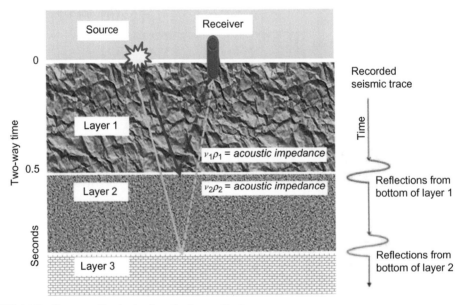

FIGURE 14.7 Conceptualized seismic reflection method.

FIGURE 14.8 A flat spot and different DHIs on the seismic section. *Source: www.sciencedirect.com and PGS Geophysical.*

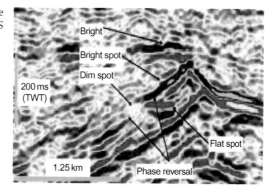

Seismic amplitude can be used to predict reservoir fluid and lithology directly from the seismic data set. For this reason, geoscientists ensure that true seismic amplitude is preserved in the seismic data during processing of the data rather than trying to increase the amplitude to maximize the structural content. By maintaining true amplitude in the seismic data, geoscientists have noticed anomalously high amplitudes or bright spots that can be equated to the qualities of oil and or gas. Such large amplitudes are a result of very large changes in density and velocity between water-saturated reservoirs, and very 'soft' oil- and gas-saturated reservoirs. The geoscientists call these and other oil or gas related anomalies direct hydrocarbon indicators (DHI). Figure 14.8 shows different DHIs on the seismic section.

RESERVOIR IMPEDANCE CONTRAST AND DIRECT HYDROCARBON INDICATORS

In seismic interpretation, the reflection coefficient (RC) of two different rocks depends on the difference between their seismic impedance and this depends mostly on the P-wave velocity in the rock. For example, the contact between sand with water saturation and sand with certain gas saturation gives a high RC that produces an amplitude anomaly in the seismic trace.

Also, the impedance contrast between reservoir sand and shale relates directly to the seismic amplitudes. Small contrasts give low amplitudes, while large contrasts cause high amplitudes.

Shale is typically harder than reservoir sand, that is, the sand is softer. Hydrocarbons soften the sand impedance even further, thus making the contrast larger.

Seismic Attribute

Seismic attribute analysis involves extracting or deriving a quantity from seismic data that can be analysed in order to enhance information that might be more subtle in a traditional seismic image, leading to a better geological or geophysical interpretation of the data. Examples of attributes that can be analysed include mean amplitude, which can lead to the delineation of bright spots, flat spots, polarity reversals and dim spots, coherency, and amplitude versus offset. Attributes that can show the presence of hydrocarbons are called direct hydrocarbon indicators (DHIs).

Direct hydrocarbon indicators (DHIs) are seismic measurements that indicate the presence of hydrocarbon accumulation in the subsurface. Types of DHI are bright spots, flat spots, 'dim spots', polarity reversals, AVO, frequency changes, etc.

Bright Spot

A bright spot is a high amplitude (high reflectivity) seismic attribute anomaly that indicates the presence of gas (or 'soft' oil) in seismic data (Figure 14.9), which has a significantly lower velocity than in a brine-saturated rock.

When water sand has lower acoustic impedance than its embedding shale (Figure 14.10), changing the water in the pores to hydrocarbon increases the sand–shale impedance contrast, and as a consequence increases the seismic reflection amplitude, which results in a bright spot.

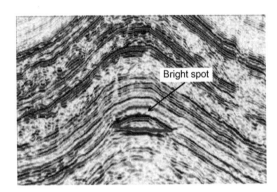

FIGURE 14.9　A 'bright spot' on the seismic section. *Source: PGS Geophysical, copied from Seismic Atlas of Southeast Asian Basin, November 2008.*

Bright spot

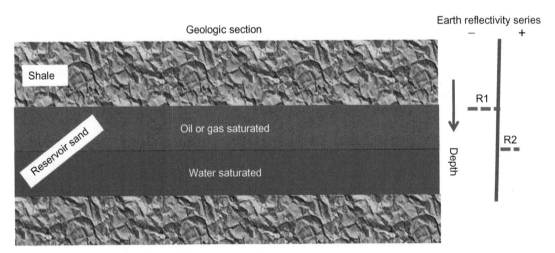

FIGURE 14.10　The effect of fluid on the earth reflectivity response.

In Figure 14.10, the shale at the top of an oil or gas sand causes the reflection coefficient to be negative. Therefore, if the water-saturated sand has a lower acoustic impedance than the overlying shale (Figure 14.11), the presence of oil or gas will make it even softer. On the model seismic response in Figure 14.12, you will notice an increased amplitude (increased negative reflection coefficient) over the crest of the structure where the hydrocarbon is present. This is called the 'bright spot'.

A seismic wave travels slower in oil than in pore water, but the difference is not always enough for a bright spot to be noticeable on the seismic section. Luckily, where there is oil, there is usually either saturated gas or gas on top, so bright spots are useful in oil exploration.

Note that seismic amplitude is directly proportional to reflectivity.

Also note that amplitude anomalies related to hydrocarbons can be expected as an on-structure brightening, as seen in Figure 14.12. The fluid contact area is identified by a sharp structurally conformable decrease in amplitudes in the off-structure direction, as seen in Figure 14.12. A conformable amplitude fit to structure can be a significant positive indicator of the presence of hydrocarbons.

Caution: Not all 'bright spots' are caused by the presence of hydrocarbon because an increase in the acoustic impedance of water-saturated sand can also be caused by an increase

FIGURE 14.11   Reflection response of a 'bright spot'.

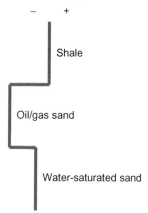

FIGURE 14.12   Model of a bright spot response, after M. Bacon et al. (2003), 3-D seismic interpretation.

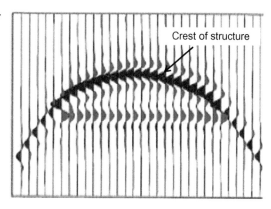

in porosity. Therefore, the 'bright spot' we notice on the seismic section should conform to structure; in map view, the amplitude changes should follow a depth contour.

Flat Spot

A 'flat spot' is a seismic anomaly that appears as a horizontal reflector on a seismic section. Flat spots will occur when there is a contact between oil, gas and water in a limited area and the surrounding reflectors are not flat. The flat spot in itself is not an amplitude anomaly, but since the contact is often associated with gas, the area can contain a bright spot (Figure 14.14).

'Flat spots' can occur when hydrocarbon-saturated sand with a lower acoustic impedance overlies water-saturated sand with a higher acoustic impedance (Figure 14.13). The 'flat spot' is noticed at the hydrocarbon–water contact. This is always a hard reflector, with a positive reflection coefficient. The flat positive reflection is visible on a seismic section because it is 'flat' in depth and will contrast with the surrounding dipping reflections (Figure 14.14).

Gas–oil contact (GOC) and oil–water contact (OWC) always have positive reflection coefficients. This is shown in Figure 14.15.

FIGURE 14.13   Acoustic impedance response of a 'bright spot'.

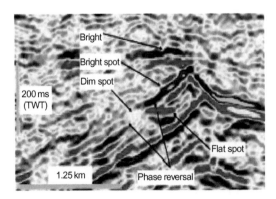

FIGURE 14.14   A flat spot and different DHIs on the seismic section. *Source: www.sciencedirect.com and PGS Geophysical.*

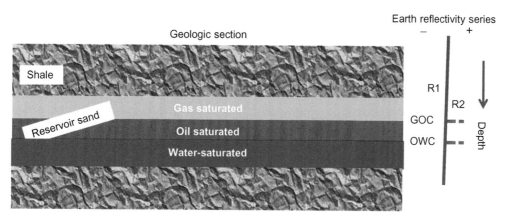

FIGURE 14.15 The effect of fluid on the reflection response that causes two 'flat spots' to appear in a seismic section.

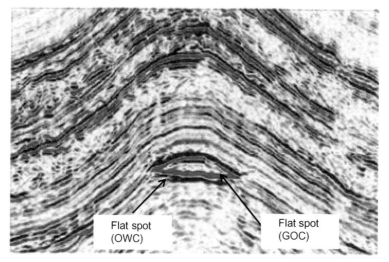

FIGURE 14.16 Seismic section showing two 'flat spots'. That is, gas–oil contact and oil–water contact. *Source: PGS Geophysical, copied from Seismic Atlas of Southeast Asian Basin, November 2008.*

Note that on a seismic section, the top of sand corresponds to a trough while the base of sand corresponds to a peak.

Note also that if the reservoir sand contains both oil and gas, then we may see two 'flat spots'. One of the 'flat spots' would be at the gas–oil contact (GOC) and the other at the oil–water contact (OWC). The 'flat spot' would be very bright (sharp amplitude) in the gas–oil contact compared to that in the oil–water contact. This is shown in Figure 14.16.

Caution: Mineralogical change and residual low-saturation hydrocarbons in the subsurface or unresolved shallower multiple reflections can cause apparent 'flat spots' on a seismic section. It is therefore important to interpret a 'flat spot' after depth conversion to confirm that the anomaly is actually flat.

Polarity Reversal

A polarity reversal is a seismic amplitude anomaly that can indicate the presence of hydrocarbon in a seismic section. Polarity reversals occur when water-saturated sand has a higher acoustic impedance than the overlying shale (shale has a lower acoustic impedance as shown in Figure 14.19), but hydrocarbon sand is softer (has a lower acoustic impedance than both the water-saturated sand and the overlying shale) (Figures 14.17–14.19).

The top sand will have a higher acoustic impedance (hard loop) below the fluid contact (top of the water-saturated sand) and low acoustic impedance (soft loop) above it, with polarity change at the contact. This change in acoustic impedance from an increase to a decrease results in polarity of the seismic response being reversed as opposed to the normal SEG convention.

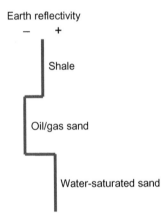

FIGURE 14.17   Reflection response of 'polarity reversal'.

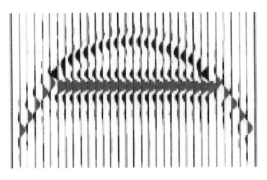

FIGURE 14.18   Model of a 'polarity reversal' response, after M. Bacon et al. (2003), 3-D seismic interpretation.

FIGURE 14.19 Acoustic impedance response of 'polarity reversal'.

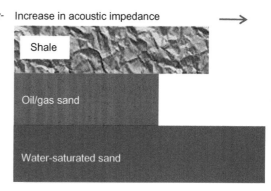

For a polarity reversal to occur, the shale has to have a lower acoustic impedance than the water-saturated sand, and both are required to have a higher acoustic impedance than oil/gas sand.

Note that compaction causes the acoustic impedance of sands and shales to increase with age and depth, but it does not happen uniformly. For younger shallow clastic rocks, shales have a higher acoustic impedance than younger sands, but this reverses at depth; for deeper clastic rocks, sands have a higher acoustic impedance than older shales.

Figure 14.20 (right) are the generalized curves showing how the acoustic impedances of gas sands, water sands and shales increase with depth.

Bright spots occur above depth A, where there is a large contrast in shale and gas–sand impedances but a modest difference between shale and water–sand impedances.

Polarity reversals occur between depth A and B, where the water–sand impedance is greater than shale impedance, but the gas–sand impedance is less than shale impedance.

Dim spots occur between depth B, where the three impedance curves converge and there is only a small impedance contrast between shale and either type of sand, brine-filled or gas-filled.

FIGURE 14.20 Normal compaction curve. *Source: Dim Spots in Seismic Images as a Hydrocarbon Indicators by Alistar R. Brown.*

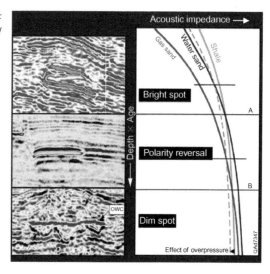

The dim spot shows a strong oil–water contact reflection, but the reflection from the top of the oil sand is of low amplitude and difficult to see because it is a dim spot.

Left of Figure 14.20, are examples of seismic reflectivity for each of the three sand/shale impedance regimes, taken from dim spots in seismic images as hydrocarbon indicators.

Dim Spot

A 'dim spot' is a low amplitude seismic anomaly that can indicate the presence of hydrocarbons. A 'dim spot' results from a decrease in acoustic impedance when hydrocarbon, which has a low acoustic impedance, replaces the water in a porous reservoir rock, which has a higher acoustic impedance than its embedding shale (Figures 14.21–14.23).

Note that not all dim spots are caused by the presence of hydrocarbons.

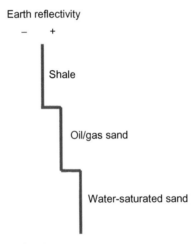

FIGURE 14.21   Reflection response of a 'dim spot'.

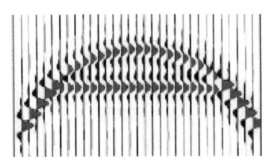

FIGURE 14.22   Model of a 'dim spot' response, after M. Bacon et al. (2003), 3-D seismic interpretation.

FIGURE 14.23 Acoustic impedance response of a 'dim spot'.

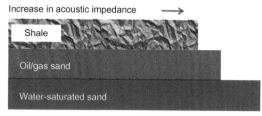

PROPAGATING SEISMIC VELOCITY VALUES IN MEDIA

Theoretical values of propagating velocities in different media:

Air	350 m/s
Weathered surface	400–600 m/s
Gravel, rubble and dry sand	400–900 m/s
Clay	900–2700 m/s
Freshwater	1400–1500 m/s
Seawater	1460–1530 m/s
Shale	1500–4000 m/s
Sandstone	1850–5200 m/s
Chalk	1800–4000 m/s
Limestone	2100–6000 m/s
Dolomite	4000–7000 m/s
Rock salt	4300–5200 m/s
Granite	4500–5800 m/s
Metamorphic rock	3000–7000 m/s
Ice	3200 m/s

Further Reading

Aki, K., Richards, P.G., 1980. Quantitative Seismology, Theory and Methods, vol. i and ii. W.H. Freeman, San Francisco.

Bacon, M., Simm, R., Redshaw, T., 2003. 3–D Seismic Interpretation. Cambridge University Press. ISBN 0-521-79203-7.

Brown, A.R., 1999. Interpretation of Three-Dimensional Seismic Data, fifth ed. American Association of Petroleum Geologists, Tulsa.

Brown, A.R., 2004. Reservoir Identification. AAPG Memoir 42 and SEG Investigations in Geophysics, No. 9 Chapter 5. pp. 153–197.

Brown, A.R., 2010. Dim Spots in Seismic Images as Hydrocarbon Indicators. AAPG Search and Discovery Article #40514. http://www.searchanddiscovery.com/documents/2010/40514Brown/.

Brown, A.R., 04 July 2012. Interpretation of Three-Dimensional Seismic Data, sixth ed. (AAPG Memoir/SEG Investigations in Geophysics No. 9). ISBN 0891813640.

Castagna, J.P., Batzle, M.L., Kan, T.K., 1993. Rock physics – the link between rock properties and AVO response. In: Castagna, J.P., Backus, M. (Eds.), Offset-Dependent Reflectivity – Theory and Practice of AVO Analysis. Investigations in Geophysics No. 8, Society of Exploration Geophysicists, Tulsa, pp. 135–171.

Coe, A.L. (Ed.), 2003. The Sedimentary Record of Sea-Level Change. Co-published by The Open University and Cambridge University Press, 288 pages.

Fazli Khani, H., Back, S., 2012. Temporal and lateral variation in development of growth fault and growth strata in Western Niger Delta, Nigeria. AAPG Bull. 96, 595–614. www.bakerhughes.com/.../well-correlation. www.sub surfwiki.org/wiki/horizon.

Gluyas, J., Swarbrick, R., September 2003. Petroleum Geoscience. Willey-Blackwell Publishing. ISBN 978-0-632-03767-4.

Greenlee, S.M., Gaskins, G.M., John, M.G., July 1994. 3-D seismic benefits from exploration through development: an Exxon perspective. Leading Edge 13, 730–734.

Hall, M., Trouillot, E., 2004. Predicting stratigraphy with spectral decomposition. In: Canadian Society of Exploration Geophysicists annual conference, Calgary, May 2004. http://www.glossary.oilfield.slb.com/Display.cfm?Term=dim%20spot http://www.slb.com/.../well_correlation.

http://Support.roxar.com/well-correlation.

Løseth, H., et al., 2009. Hydrocarbon leakage interpreted on seismic data. Mar. Petrol. Geol. 26 (7), 1304–1319.

Mariller, F., Eichenberger, U., Sommmaruga, A., 2006. Seismic Synthesis of the Swiss Molasse Basin. Institute de Geophysique, UNI Lausanne, CPI, 1015 – Lausanne.

Petrel Seismic Attribute Analysis. Schlumberger. http://www.slb.com/services/software/geo/petrel/seismic/seismic_multitrace_attributes.aspx.

Shuey, R.T., 1985. A simplification of the Zoeppritz equations. Geophysics. 50, 609–614.

Simm, R., White, R., 2002. Phase, polarity and interpreter's wavelet. First Break. 20, 277–281.

Index

Note: Page numbers followed by *f* indicate figures and *t* indicate tables.

08/05/2025

01864895-0001